Network Science

Francesca Biagini · Göran Kauermann ·
Thilo Meyer-Brandis
Editors

Network Science

An Aerial View

Editors
Francesca Biagini
Department of Mathematics
LMU Munich, Munich, Germany

Göran Kauermann
Department of Statistics
LMU Munich, Munich, Germany

Thilo Meyer-Brandis
Department of Mathematics
LMU Munich, Munich, Germany

ISBN 978-3-030-26816-9   ISBN 978-3-030-26814-5   (eBook)
https://doi.org/10.1007/978-3-030-26814-5

This Springer imprint is published by the registered company Springer Nature Switzerland AG
The registered company address is: Gewerbestrasse 11, 6330 Cham, Switzerland

Preface

The book was initiated through a joint research focus on quantitative network science funded by the Center for Advanced Studies at the Ludwig-Maximilans-University Munich (LMU) from 2015 to 2017. A group of researchers from different scientific fields launched the quantitative network science center at LMU which included different scientific disciplines that dealt with network data in the widest sense. A final workshop in October 2016 covered the selected fields of the center with its multiple perspectives on network science. Some of these fields are mirrored in this book. First, networks can be treated as a mathematical object which can be analyzed from different perspectives. Secondly, networks can be considered as data which need to be analyzed. This requires visualization tools as well as methods and models which allow to draw conclusions from the network data. Finally, a network itself can be considered as a model where the entities interact with each other based on the model structure. How is information transmitted through the network? How does the network behave if shocks occur or if certain nodes fail? Such questions relate to transmission of diseases but also to investigation of systemic risk in financial networks. Sometimes the network structure itself is the focus of interest. If the links between edges are unknown, research questions can focus on finding the driving network structure. Such questions are central in genetic networks, where inference about the network structure is drawn from data. This also applies to extreme events in a network. And finally, networks are per se visualizations, since a network as object is just a matrix. The different aspects sketched above are tackled in the chapters of this book.

The editors would like to thank the Center for Advanced Studies at the Ludwig-Maximilans-University Munich for funding a research group in the field of network science. This has led to the above mentioned workshop and subsequent cross-discipline collaboration. We would also like to thank the European Cooperation in Science and Technology (COST action CA15109—COSTNET), which supported some of the contributions in this book. Last, but most prominently, we do however want to thank the authors of this volume for their contributions and the hard work they put in it. Finally, tremendous credits need to go to Michael Lebacher, who read several chapters of the book carefully and provided extremely helpful remarks to the authors. Many thanks!

Munich   Francesca Biagini
March 2019   Göran Kauermann
                                                                               Thilo Meyer-Brandis

Contents

Contributors

Francesca Biagini Department of Mathematics, University of Munich, Munich, Germany

Ulrik Brandes Department of Humanities, Social and Political Sciences, ETH Zurich, Zurich, Switzerland

Steffen Dereich Institute of Mathematical Stochastics, Westfälische-Wilhelms Universität Münster, Münster, Germany

Nils Detering Department of Statistics and Applied Probability, University of California, California, USA

David R. Hunter Department of Statistics, Penn State University, State College, USA

Göran Kauermann Department of Statistics, University of Munich, Munich, Germany

Claudia Klüppelberg Center for Mathematical Sciences, Technical University of Munich, Munich, Germany

Steffen Lauritzen Department of Mathematical Sciences, University of Copenhagen, Copenhagen, Denmark

Thilo Meyer-Brandis Department of Mathematics, University of Munich, Munich, Germany

Konstantinos Panagiotou Department of Mathematics, University of Munich, Munich, Germany

Daniel Ritter Department of Mathematics, University of Munich, Munich, Germany

Michael Sedlmair Department of Computer Science, University of Stuttgart, Stuttgart, Germany

Ernst C. Wit Institute of Computational Science, Università Della Svizzera Italiana, Lugano, Switzerland

Chapter 1
Introduction

Francesca Biagini, Göran Kauermann and Thilo Meyer-Brandis

Abstract Network Science is a term used for a wide field of methods all related to analyzing networks and/or network data. This ranges from mathematical questions to applied data analytic problems. We give a general overview of the different aspects and refer to the chapters in this book.

Network science, the science of analyzing networks, has become increasingly important. A network is a collection of actors (nodes), which are connected with each other (through edges). Examples include genetic, social, and traffic networks, to name but a few. Research questions are, among others, the dynamic behavior of the network, the transmission of information through the network, or the network structure itself. Networks are simple in structure and in principle one can even represent the network as squared matrix, so that edges are numbers in the matrix. A friendship network can for instance be written as matrix with entries 1 (for an existing edge) and 0 (otherwise). And while the structure of a network is simple, its analysis and modeling can get rather complex. Moreover, if a network gets large, the behavior in the network follows asymptotic rules, whose derivation is challenging. These aspects become apparent with the increasing availability of network data. Today, we live in a data-driven society in which information is measured, recorded, and stored in many areas of daily life, and network data, available in nearly all of these areas, need to be analyzed.

Network science and network data analysis is not confined to any single scientific discipline but similar quantitative needs arise in diverse fields such as industry, engineering, genetics, medicine, biology, economics, and social sciences, among many

F. Biagini · T. Meyer-Brandis
Department of Mathematics, University of Munich, Munich, Germany
e-mail: biagini@mathematik.uni-muenchen.de

T. Meyer-Brandis
e-mail: meyer-brandis@math.lmu.de

G. Kauermann (✉)
Department of Statistics, University of Munich, Munich, Germany
e-mail: goeran.kauermann@stat.uni-muenchen.de

© Springer Nature Switzerland AG 2019
F. Biagini et al. (eds.), *Network Science,* https://doi.org/10.1007/978-3-030-26814-5_1

others. Network science is therefore strongly interdisciplinary and includes methods and theories ranging from mathematical graph theory and statistical network models to visualization techniques in computer science. The contributions in this book mirror this wide range and illustrate the different research questions being tackled. Network science has attracted more and more interest since the 1960s and has grown significantly in recent years. Networks provide an abstract way of describing relationships and interaction between elements of complex and heterogeneous systems.

Since networks can be represented as graphs, network science is strongly influenced by graph theoretical approaches. Even though networks can be applied to different fields, they nonetheless share the common structure that actors (nodes) build up ties (edges) to other actors in the network. The complexity of network analyses can grow rapidly. First, networks can be dynamic, i.e., the network structure changes over time. Second, the nodes in the network can change or be heterogeneous. Third, the links between the nodes can be complex and multitude. In general, networks carry complex interactions, and network models are essential for analyzing, visualizing, and understanding activities in the network.

The aims of network analyses are multifold, using theoretical, numerical, and data analytic tools. One can, for example, investigate how a network behaves in extreme cases, e.g., when the number of actors (nodes) is growing, by developing suitable mathematical models. Numerical methods can be employed to find the shortest paths between two actors in a network, while statistical methods may be used to model time dependence and changing network structures. Networks itself occur in nearly all fields. For example, the World Wide Web can be represented as a network whose vertices are HTML documents, connected by the hyperlinks that point from one page to another. On a different level, our nervous system forms a large network, whose vertices are the neurons and nerve cells, which are connected by axons. Complex networks are also considered in social and economic sciences. Here the vertices represent (specific groups of) individuals or entities, and the edges describe social or some other type of interaction between them. Yet another example of the use of networks is in information visualization and visual analytics in order to discover unexpected patterns in network data.

In fact, the field of network science is too broad to cover it comprehensively in a single book. We therefore do not intend to give a full picture of network science but to illuminate some aspects in the area which are current active research fields. The contributions in the book provide an aerial view and by doing so the reader can see the wide angle of network-related research.

Any analysis of networks should start by visualizing them. While networks are mathematically clearly defined objects, which are either represented by a set of nodes and a set of edges, or equivalently by an adjacency matrix, their visualization is in no way unique or even simple. Chapter 2 presents the state of the art in this field. At first sight, it does not appear to be difficult to visualize networks. Nodes are drawn with circles and edges are drawn as links between them. While this is certainly true, the question remains how to allocate the nodes and how to draw the edges when visualizing the network. Network maps of underground trains can serve as a simple example. They fulfill the purpose of easy readability but cannot serve as a map of a city. While we are all used to read such network graphs, we need to question what is intended to be visualized and how can this be done if networks grow in size. The hairball effect is one issue, and computation time is another constraint. An alternative representation results through comprehending the network as a matrix and directly visualizing the matrix. The second chapter of this book gives some ideas in this direction by focusing on some network examples, which are represented in different styles. This mirrors the complexity of the field but also shows some novel ideas.

If the network is considered as adjacency matrix, one may treat this quantity as multivariate random variable. In other words, edges between the actors result from a random process, which itself might be influenced by the network. This view leads to statistical network data analysis which is portrayed in Chap. 3. The central model in this field is the so-called exponential random graph model (ERGM), which states that the network and its distribution are described by a number of quantities calculated from the network. Such quantities, usually called statistics, are for instance the number of edges, the number of the so-called two stars (or "V" constellation, i.e., three nodes connected in form of a "V"). While the models are statistically appealing, their estimation based on data is numerically demanding. A second strand of models results by focusing on changes in a network. If the network is not considered as static but dynamic in that edges may develop or disappear, one can consider this as a stochastic process. In principle, one could comprehend the resulting data as a time series of networks. Chapter 3 sketches the up-to-date statistical approaches in this field and closes with a different statistical field using networks, namely graphical models. The latter is picked up again in Chap. 6.

With the birth of Internet-based social networks, information on personal interaction on a large-scale became available and the interest in developing mathematical models that are able to capture and explain certain stylized features of large and complex real-world networks has grown significantly. Among the most central phenomena observed in large-scale real-world networks is the so-called small-world phenomenon, which refers to the fact that despite the large network size most of the nodes have a surprisingly small graph (hop count) distance in between them, and the so-called scale-free nature of networks, meaning that the degree sequence has polynomial decay. Chapter 4 focuses on two paradigms of random graph models that have been thoroughly studied over recent years, the rank-one and the preferential attachment paradigm. More precisely, it introduces certain random graph examples that are archetypical for the rank-one and preferential attachment model classes and illustrates how the above-mentioned phenomena are addressed and studied within these types of model classes.

In an ever more connected world, the notion of systemic risk, which can be described as the risk that a local shock to a system spreads to substantial parts of the system due to contagion or infection effects, becomes a central concern. Examples include epidemic spreads of diseases, rumors spreading in social networks, breakdowns of power grids, or the collapse of financial networks. One classical contagion process studied in the framework of random graphs is the so-called bootstrap percolation where a node gets infected as soon as a fixed number of neighbors are infected. However, both the classical bootstrap percolation process and the homogeneous Erdös–Rényi graph do not account for the strong heterogeneity observed in most real-world networks. Chapter 5 summarizes some recent results on how a generalized bootstrap percolation process can be studied on inhomogeneous random graphs with the focus on measurement and management of systemic risk caused by default contagion in financial networks. In addition to quantifying the impact of a local shock, resilience criteria and sufficient capital requirements to stabilize a given system are obtained in terms of network characteristics.

Chapter 6 deals with the question of how to model and analyze extreme risk that evolves in some system according to certain network dependencies. Consider, for example, the event of extreme rainfall on a specific location within a network of rivers. This event will affect water levels at other parts of the system in a specific way depending on the precise structure of the river network. In Chap. 6, Bayesian networks associated with directed acyclic graphs are used to model this transmission of extreme risk, where the conditional distribution of the state of a given node is described in terms of the distribution of the states of its parental nodes, i.e., the nodes in the network pointing toward the given node. In this sense, Bayesian network models are closely related to vector autoregressive models. In case of extreme risk events, the involved distributions are typically heavy-tailed and the conditional distributions can be described by the so-called max-linear structural equations. Within this framework, independence properties of the model and parameter estimation are discussed.

In genomics, language and tools from network science have been actively employed to study complex genetic systems. Roughly speaking, nodes in a genetic network are abstractly described as "genes," and edges between the genes represent some kind of "genetic interactions." Since in genetic networks the (random) process of interest typically lives on the nodes and not on the edges, the study of genetic networks mostly involves dynamical system models used to describe flows rather than random graph models as described in Chaps. 4 and 5. Chapter 7 gives an overview of three common types of models used in the study of genetic networks depending on the genomic network data and experiment under consideration. More precisely, to study the so-called mechanistic genomic networks of molecular interactions within a single cell, a system of stochastic differential equations is proposed that takes into account the underlying stochasticity of molecular interactions. To model the evolution of the so-called functional genomic networks where the focus is on systems on a larger scale such as organs or other biological subsystems, systems of ordinary differential equations are more appropriate. Finally, vector autoregressive models show to be useful as an alternative for large genomic systems, where systems of (stochastic) differential equations might become unstable and computationally prohibitive.

The authors of the chapters come from different scientific areas, make use of different methods, investigate different research questions, and apply different tools. Still, they have one thing in common. They all consider networks and network data and by doing so, the contributions in this book are apparently tied together. This combination of different aspects of network science provides the intended aerial view.

Chapter 2
Network Visualization

Ulrik Brandes and Michael Sedlmair

Abstract Data visualization is the art and science of mapping data to graphical variables. In this context, networks give rise to unique difficulties because of inherent dependencies among their elements. We provide a high-level overview of the main challenges and common techniques to address them. They are illustrated with examples from two application domains, social networks and automotive engineering. The chapter concludes with opportunities for future work in network visualization.

2.1 Introduction

Data visualization is the art and science of mapping data to graphical variables in a way that facilitates the identification of individual values and aggregate patterns. The main motives are data exploration for analysts and communication of information toward a recipient. One should not underestimate, however, some of the more circumstantial aspects of visualization: decorative appeal, symbolism, and suggestiveness. Classics include Bertin (1983) and Tufte (1983); for recent research-oriented overviews, see, for instance, Munzner (2014) and Grant (2018).

Networks, as a special form of data, pose unique challenges to visualization because of inherent trade-offs and dependencies among the elements in a graphical mapping. A network diagram of a subway system, for instance, should facilitate travel planning so that finding stations and following lines takes precedence over topographic accuracy. Since Harry Beck designed the first schematic map of the entire London Underground network in 1933, it has become common to restrict the slopes of lines which, in turn, also influence where stations can be placed. While there are many aspects to schematic map design (Roberts 2012), we illustrate in Fig. 2.1 the effect of network schematization on the representation of distances.

U. Brandes (✉)
Department of Humanities, Social and Political Sciences, ETH Zurich, Zurich, Switzerland
e-mail: ubrandes@ethz.ch

M. Sedlmair
Department of Computer Science, University of Stuttgart, Stuttgart, Germany
e-mail: michael.sedlmair@visus.uni-stuttgart.de

© Springer Nature Switzerland AG 2019
F. Biagini et al. (eds.), *Network Science*, https://doi.org/10.1007/978-3-030-26814-5_2

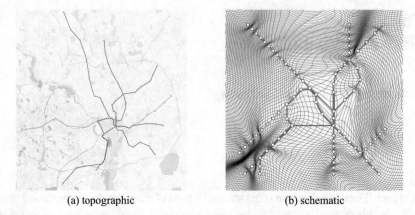

(a) topographic   (b) schematic

Fig. 2.1 A topographically accurate and a schematic version of the Washington Metro system. As is common for metro maps, the schematic version exhibits restricted angles, an enlarged center, and equidistant stations along arteries. The underlying grid indicates the distortion of actual spatial relationships. (Adapted from Boettger et al. 2008. Published with permission of © IEEE 2008)

Layout problems become even more challenging when networks are large. In addition to design criteria and visual clutter, for instance, computational complexity of layout algorithms is a concern. Many other challenges beyond layout need to be addressed, including multivariate information associated with nodes and links and alternative representations such as adjacency matrices.

Altogether, this makes network visualization a highly interesting design problem in which many trade-offs and interdependencies among aspects as diverse as traditions, aesthetics, and constraints, as well as computational issues need to be taken into account.

Our contribution to this volume is split into two main parts. The first part, in Sect. 2.2, provides a high-level overview of the main challenges associated with the visualization of networks and a glimpse at some of the more common techniques to address them. In Sect. 2.3, we look more concretely at the visualization of networks from two different applications domains. The chapter concludes with opportunities for future work in network visualization.

2.2 Principles

Adopting the notation used throughout this book, we define a (binary) network with n nodes as a (binary) data matrix $y \in \{0, 1\}^{n \times n}$. More general situations are considered briefly in Subsect. 2.2.4.

Network visualizations are commonly produced using techniques developed for graphs. A graph representation $G(y) = (V, E)$ of a network y consists of the set $V = \{1, \ldots, n\}$ of vertices and the set $E = \{(i, j) \in V \times V : y_{ij} = 1\}$ of (directed)

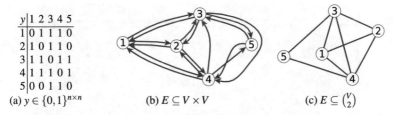

$$\begin{array}{c|ccccc}
y & 1 & 2 & 3 & 4 & 5 \\ \hline
1 & 0 & 1 & 1 & 1 & 0 \\
2 & 1 & 0 & 1 & 1 & 0 \\
3 & 1 & 1 & 0 & 1 & 1 \\
4 & 1 & 1 & 1 & 0 & 1 \\
5 & 0 & 0 & 1 & 1 & 0
\end{array}$$

(a) $y \in \{0,1\}^{n \times n}$ (b) $E \subseteq V \times V$ (c) $E \subseteq \binom{V}{2}$

Fig. 2.2 A network y and visualizations of the associated directed and undirected graphs $G(y) = (V, E)$ with different layouts and edge styles

edges. If y is symmetric and non-reflexive, i.e., $y_{ij} = y_{ji}$ and $y_{ii} = 0$ for all $1 \le i, j \le n$, the network can also be represented as an undirected graph in which the edges are unordered pairs $\{i, j\} \in E \subseteq \binom{V}{2}$. Sometimes graphs are mixed, with both directed and undirected edges.

Multiple forms of visualization have been devised for graphs. In the first subsection below, we introduce the most typical and intuitive design, node-link diagrams, that is exemplified in Fig. 2.2. It is by far the most frequently used, and since its spatial arrangement has more degrees of freedom than a statistical chart, we briefly review the problem of layout in Subsect. 2.2.2. In Subsect. 2.2.3, we give special consideration to large networks. This section ends with a discussion of the visualization of multivariate network data and some ideas on alternative representations.

2.2.1 Standard Representation

The most common representation of graphs is the node-link diagram in which vertices are represented by point-like features and edges by curves or line segments connecting them. Examples are shown in Fig. 2.2 and alternatives are discussed in Subsect. 2.2.5.

Node-link diagrams appear long before Euler's seminal work on the bridges of Königsberg, which is usually considered the beginning of graph theory but does not contain any drawing of a graph. For centuries, point features and connecting curves had already been used on maps and for non-spatial relations, including ancient board games, astrological and logical diagrams, ancestral relations, and geometric figures. Kruja et al. (2001) give a short history of graph drawings, although many more stunning examples such as the one shown in Fig. 2.3 exist.

Although the standard representation is intuitive, it is prone to a number of potential shortcomings. Experiments show that small angles and crossings of edges may hinder reading (Purchase 2000; Ware et al. 2002); spatial proximity suggests cohesive groups even when they are connected only loosely (McGrath et al. 1996), and, quite obviously, identification of elements becomes difficult in the presence of overlap.

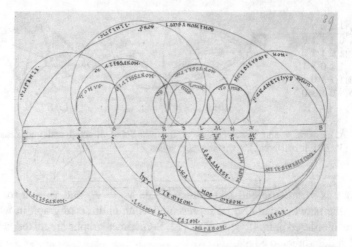

Fig. 2.3 Diagram showing musical relationships in a twelfth-century copy of a manuscript of Boëthius (c. 480–524). Source: St. Gallen, Kantonsbibliothek, Vadianische Sammlung, VadSlg Ms. 296, f. 89r. Boethius, De arithmetica, De institutione musica, https://www.e-codices.ch/en/list/one/vad/0296, under the CC BY-NC 4.0 license. Published with permission

2.2.2 Layout

The research field of graph drawing (Tamassia 2013) is concerned with geometric representations of, and layout algorithms for, graphs and hypergraphs, whereas visualization design, task appropriateness, and user interaction are more commonly studied in the areas of information visualization, visual analytics, and human–computer interaction.

Spatial arrangement, or layout, is a non-trivial issue with any graphical representation. Unlike statistical charts such as scatterplots, time series, or pie charts, however, node-link diagrams usually do not come with given relative positions and thus exhibit more degrees of freedom. This is a curse and a blessing, because on the one hand, a layout can be adjusted to express additional information and increase readability, and on the other hand, edges create complex dependencies turning both into rather daunting tasks.

Readability criteria such as density distribution, size of angles, number and angles of crossings, bends, area, symmetry, etc. are sometimes referred to as aesthetic criteria, and their priorities may be influenced by the task at hand. In graph drawing, layout algorithms take them into account as constraints or optimization objectives.

While specialized algorithms have been proposed for classes of graphs such as trees and planar graphs, and for representation variants such as layered or orthogonal layouts, one group of layout algorithms clearly dominates the practical use of algorithms for general undirected graphs. They are referred to as force-directed algorithms (Brandes 2016; Kobourov 2013) because they are inspired by physical analogies of mutually repelling nodes (for good distribution and overlap avoidance) and edges acting as springs between them (for uniform edge length and visually recognizable group cohesion). As a by-product, symmetries can often be recognized well, and the simulation of physical forces facilitates modes of interaction that feel natural to users.

On the negative side, the simple, intuitive, and widely available implementations are non-deterministic, sensitive to poor initialization, they often get stuck in local minima from which the iterative improvements strategy is unable to escape, and they have difficulties with graphs of low diameter or large size as evidenced in Fig. 2.5.

The most robust and reliable variants are instances of multidimensional scaling with the graph-theoretic distances as input and coordinates as output. Instead of a force calculation determining an update step, minimization of a layout objective function is attempted. For a two-dimensional layout $p = (p_i)_{i \in V} \in \mathbb{R}^{n \times 2}$ with $p_i = \langle x_i, y_i \rangle$, are examples. the squared relative error of shortest-path distances $dist(i, j)$ in the underlying undirected graph represented by Euclidean distances $\| p_i - p_j \|$ in the layout is defined as

$$stress(p) = \sum_{i, j \in V} \frac{1}{dist(i, j)^2} \left(\| p_i - p_j \| - dist(i, j) \right)^2.$$

Exact minimization is computationally intractable but with good initialization and carefully designed iterative improvement procedures such as majorization (Gansner et al. 2004; Wang et al. 2018), low-stress layouts can be obtained reliably and efficiently (Brandes and Pich 2008). Approximate minimization for large graphs is discussed in the next subsection.

Note that it is precisely the focus on distances that renders networks such as the one depicted in Fig. 2.5 problematic. However, the stress objective can be modified by altering the notion of target distances $dist(i, j)$, by varying the relative contribution of dyads (i, j), or by building auxiliary graph structures that may include virtual vertices and edges. We give an example in Fig. 2.6.

Other layout requirements may be expressed as constraints that restrict the space of admissible layouts for instance by fixing vertices to certain areas or relative to each other (Dwyer 2009). The approach is thus flexible and can be adapted to more different application settings than one might initially suspect, including dynamic graphs.

The relevance of the stress-minimization approach is reinforced by the fact that other important approaches turn out to be special cases. Spectral layout, where coordinates are determined from eigenvectors of the Laplacian matrix of a graph, and barycentric layout, where some vertices are fixed and the others are placed in average position of their neighbors, are examples.

A recent development are neighborhood embeddings, where distances are determined only locally and patched together. They appear to be especially suited to highlight clustering structure (Maaten and Hinton 2008; Roweis and Saul 2000).

2.2.3 Large Networks

With increasing size of a network, the problem of visualizing its graph changes until it becomes qualitatively different.

An obvious challenge for layout algorithms is running time. Without special precautions, a single iteration of a force-directed algorithm that moves every vertex only once already requires time linear in the size of the graph. Speed-up techniques attempt to reduce the number of iterations using fast methods that get the larger distances approximately right so that the iterative procedure does local adjustments only. Multilevel methods obtain suitable initializations by recursively operating on smaller graphs (Gajer and Kobourov 2002; Hachul and Jünger 2004; Hu 2006). Simpler but no less effective is the use of approximate classical scaling (Brandes and Pich 2006) a spectral decomposition method that prioritizes larger distances and requires near-linear running time. Additionally, the time spent in iterations can be reduced by coarsening the stress function and thus eliminating redundancies (Ortmann et al. 2017), or by parallelizing algorithms for GPU computation (Wang et al. 2018).

Beyond runtime, display limitations are another concern. Even with sufficient resolution to display tens of thousands of line segments for edges, it may not be possible for a human viewer to discern the details. Worse, the nature of the stress objective is such that low variance in distances leads to largely uniform vertex distribution and cluttered edges. This is sometimes referred to as the hairball problem of small-world networks and one of the reasons Fig. 2.5 appears cluttered.

Compensation techniques include pre- and postprocessing during layout generation and level-of-detail rendering of a given layout. An example of a preprocessing technique particularly suited for graphs with low variance in distances is the determination of a backbone, i.e., a subgraph induced by edges that are contained in regions of relatively high local density (Nocaj et al. 2015). Absent many shortcut edges, distances in such backbone structures are generally larger and more varied which makes their layout easier.

Edge bundling is a technique that has been used in both, pre- and postprocessing. In the most common variants (Gansner et al. 2011; Holten 2006), a given layout is modified by bundling the middle segments of edges that would run close to parallel anyway because they start and end in similar display regions.

Abstractions and simplifications can also be accomplished in graphical space, for instance by adjusting the level of detail at which a graph with a given layout is drawn (Zinsmaier et al. 2012). A more comprehensive overview is given by Landesberger et al. (2010).

2.2.4 Multivariate Networks

A network as defined above is a single variable representing relationships between entities. In realistic data-analytic scenarios, it is unlikely to be the only variable.

Often, there will be additional node-level attributes and multiple types of relations, possibly on different sets of nodes. Even more dimensions are introduced if one or more of the variables vary over time, which results in a dynamic network. Networks made up of multiple relations are often referred to as multilayer networks (Kivelä et al. 2014) and their visualization is discussed in Kerren et al. (2014).

As an example, consider the visual encoding of two node variables in coordinates. Their quantitative, ordinal, or categorical values constrain the spatial layout, and in the extreme case, node positions are fixed as in a scatterplot (Wattenberg 2006). Note that choosing such a graphical mapping favors the nodal attribute data over network structure which was the sole criterion in the layout algorithms above.

Many other ways of visually encoding multivariate networks exist. Often, additional graphical elements such as labels, colors, and glyphs for enriching nodes, line thickness, shapes, and gradients for enriching links, or additional separating lines and boundaries to compartmentalize information are added.

We give two examples of multivariate network visualizations. The first, in Fig. 2.6, is the same network of retweeting politicians shown in Fig. 2.5 but with additional data (and a different layout as explained in Sect. 2.3.1 on social networks). The second, in Fig. 2.7, shows a sequence of 15 networks (including the one from Fig. 2.4b) of ranked friendships in a matrix representation.

The more variables to display, the greater the danger of overstuffing. It is thus important to keep general visualization guidelines in mind. When using color to group nodes into categories, it is, for instance, important to realize that a viewer will only be able to distinguish six to twelve colors reliably.

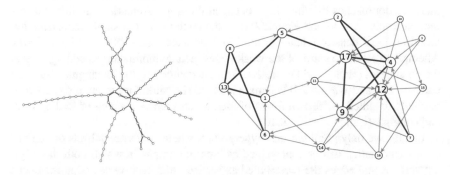

(a) Metro system treated as a network without geographical information (cf. Fig. 2.1)

b) Social network of university students (three best friends) (Newcomb 1961). See Fig. 2.7 for full data

Fig. 2.4 Network visualization using a force-directed layout algorithm

2.2.5 Other Representations

The standard representation in the form of node-link diagrams is not the only way
of visualizing networks. Straightforward variants include drawings with orthogonal
edges (as are common for circuit schematics) or other predetermined slopes (as in
metro maps). Implicit representation of edges appears, for instance, in inclusion
drawings of trees where vertices are represented as areas and these areas are placed
within other areas such that parent–child relationships can be inferred from area
inclusion.

Many other representations exist but are often feasible only for graphs that satisfy
structural properties such as acyclicity or planarity.

A common alternative that applies to general graphs is matrix representations
where rows and columns are indexed by the vertices and edges are represented in
matrix cells. A one-dimensional layout problem remains: the (joint) ordering of
rows and columns. This ordering is important as it relates to the recognizability
of structural features in the form of cell-arrangement patterns such as on- or off-
diagonal blocks (density within or between groups) and crosses (high-degree vertices
brokering between groups).

Just like graph layout algorithms, many ordering criteria and algorithms have
been considered (Behrisch et al. 2016). Most of the objectives are computationally
intractable (Díaz et al. 2002) giving rise to interesting computational challenges. An
example that includes additional aspects is given in Fig. 2.7.

A comparison between standard and matrix representations suggests that they
have complementary strengths and weaknesses (Ghoniem et al. 2004). Consequently,
there are approaches that transition between these representations at different resolution levels (Abello and Ham 2004) or combine them in a single representation
based on local density (Henry et al. 2007) or select paths (Shen and Ma 2007).
Attempts to alleviate some of the weaknesses include modifications adding cues at
the boundaries (Henry and Fekete 2007) and decomposing the rectangular area into
patches (Bae and Watson 2011; Dinkla et al. 2012). Matrix representations are also
particularly suited for certain forms of interaction such as resizing or folding rows
and columns (Elmqvist et al. 2010).

Finally, we only point out that hypergraphs, where edges are subsets of vertices
of any cardinality, can be represented as bipartite graphs in which both the original vertices and edges are represented as vertices, and each vertex-edge incidence
is represented by an edge. While this representation allows to use common graph
visualization techniques, more specific representations such as Venn diagrams exist
as well.

2.3   Practice

From a data visualization perspective, the graphical representation of a network should be designed such that relevant information can be perceived with ease and accuracy. Since data, information, and tasks differ across application domains, so does appropriate visualization. Domain traditions and prior knowledge of recipients require further adaptation.

We next discuss network visualization in two very different scenarios to illustrate the breadth and depth of issues arising.

2.3.1   Social Networks

When social phenomena are described as networks of social relations, the information to be conveyed in their visualizations may be manifold, with different foci, different aspects, and on different levels (Borgatti et al. 2018; Hennig et al. 2012). Thus, an especially rich set of tasks and visualization techniques has evolved around the concept of social networks (Brandes et al. 2013).

On the macrolevel, the interest is generally in characteristics of the social network as a whole. Such characteristics may include whether the network consists of a dense core and a loosely connected periphery or whether it is polycentric, whether there are many shortcuts that accelerate diffusion or whether there are bottlenecks, and whether subnetworks are organized hierarchically or whether the network is flat.

Certain characteristics that are commonly encountered in social networks require adaptation of layout algorithms. Networks with low average distance are known as small-world networks and an example was given in Fig. 2.5. The alternative layout for the same network in Fig. 2.6 uses a preprocessing technique that ignores edges potentially bridging different regions of the graph in the computation of shortest-path distances (Nocaj et al. 2015). These are identified as edges not embedded in a tightly knit neighborhood. Since the Twitter accounts are represented by rectangles of varying size, an additional postprocessing step is applied to reduce node overlap (van Garderen et al. 2017).

On the microlevel, the interest is in individual differences and special configurations such as node centrality and the prevalence of substructures which are sometimes referred to as motifs. Network-analytic techniques focusing on characteristics of nodes or links typically yield additional attribute data and therefore lead to multivariate network visualization problems.

It is thus typical for social networks that some data, such as party affiliation or the number of retweets on an edge, is given, and other data, such as the invisibly used status of being a backbone edge or the depicted number of retweeters, is derived from the structure.

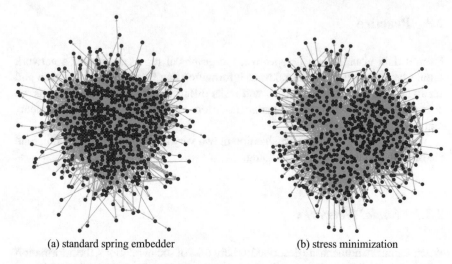

(a) standard spring embedder   (b) stress minimization

Fig. 2.5 Retweeting among 661 Twitter accounts of German politicians during the 2013 federal elections. Data collected by Lietz, Wagner, Bleier, and Strohmaier at GESIS (Lietz et al. 2014). Due to small distances, neither layout method does particularly well, although stress minimization hints at variation in local cohesion. This network is redrawn in Fig. 2.6

Longitudinal social networks arise from multiple waves of observation and are often represented using small multiples or animation (Brandes et al. 2012). A different representation is shown in Fig. 2.7. Here, a group of 17 university students has been asked repeatedly to rank their peers in order to understand the evolution of their social relationships (Newcomb 1961). While Fig. 2.4b shows just one of the resulting 15 networks, with edges only for the top three nominations of each respondent, the matrix representation in Fig. 2.7 shows, for every dyad, how the relationship develops over time: a line segment extends to the left and right according to the rank assigned to the column actor by the row actor and vice versa at a particular point of observation. The higher the rank, the longer the line, and the more imbalanced the two nominations, the more the line tilts toward the longer side. The 15 pairs of observations are vertically aligned to ease identification of trends and outliers.

In the study of ego networks (Perry et al. 2018), actors are characterized by their individual social environment. Instead of an ordered sequence of snapshots of an evolving networks, these collections represent an ensemble of networks made up of the same variables. For visual comparison, they can be arranged by similarity and summarized in subgroups of ego networks or by composition (Brandes et al. 2008).

As these few aspects indicate, social networks pose a broad range of challenges for network visualization both in terms of the richness of data and the analytic interest taken in them.

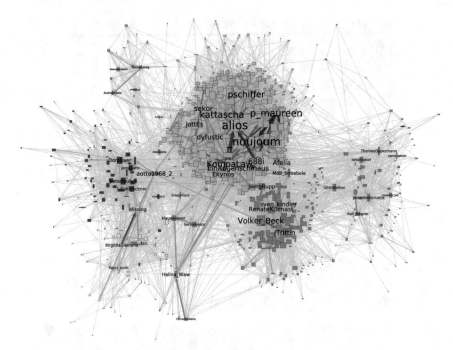

Fig. 2.6 Network of Fig. 2.5 with different layout and multivariate information. Nodes are colored according to political party affiliation, and their width and height indicate the number of others they retweet (outdegree) and have been retweeted by (indegree). Node labels are shown only for some particularly involved accounts and are scaled according to activity level (sum of in- and outdegree). Both thickness and darkness of lines indicate the number of retweets making up that particular relationship. Width, height, and thickness are all proportional to the square root of the respective quantity. The layout is determined by stress minimization of distances in a backbone structure consisting of edges in dense regions and postprocessed to reduce node overlap

2.3.2   Overlay Networks for Automotive Engineers

In the second example, we describe a particular application in more detail. A network analysis tool called *RelEx* (Sedlmair et al. 2012), which was built to support automotive engineers in understanding in-car communication networks.

In-car communication networks describe the networks that connect the electronic components of a car through the respective communication bus systems. The challenges of visualizing such networks are very different from those of analyzing social networks. Instead of the node scalability issue in social networks, the main complexity stems from the interplay of different network types that form an overlay network. While there are only a few nodes (up to 100), the network is very dense, which makes a matrix representation a viable design choice.

Fig. 2.7 Evolution of a social network of co-habitating students known as Newcomb's Fraternity (Newcomb 1961). Mutual rankings over 15 waves of observation have been stacked from bottom to top in each cell (Brandes ans Nick 2011, published with permission of © IEEE 2011). Length and tilt of each line indicate how the two students involved ranked each other. Color on the diagonal indicates the degree to which an individual is ranked above or below unbiased expectation. Rows and columns have been ordered to keep strong relationships close, resulting in two diagonal blocks that are relatively stable over time

Figure 2.8 shows a screenshot of the RelEx tool. As there are many visual analysis tools, RelEx uses multiple coordinated views. All views are interactively connected by linking and brushing, that is, selecting nodes/edges in one view gets propagated by highlighting the same nodes/edges in other views. On the top left, the physical network is shown in an automotive-typical node-link diagram, which schematically shows how all the components are wired up in the car. Abstractly, this network is an undirected hypergraph with edges that connect multiple nodes at ones. The bottom left shows the specific path a selected signal takes over this network. It is basically a filtered version of the view above, which can be interactively set by the user.

The view we want to primarily focus on is the matrix view on the right side. The matrix view provides an overview over the logical network. The logical network specifies how signals (up to 10.000) are exchanged between the hardware components. Abstractly, this network forms a directed multigraph, that is, it can have multiple edges between nodes (i.e., two components can exchange multiple different signals). We remember that the nodes in the matrix representation are now shown as the lines

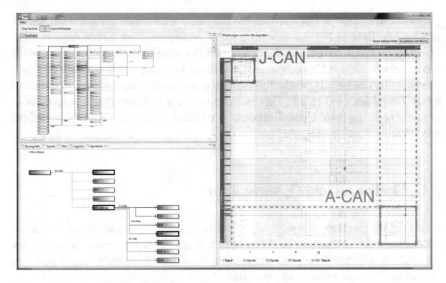

Fig. 2.8 Screenshot of the RelEx (Relation Explorer) tool (Sedlmair et al. 2012). The tool comprises three main views: (top left) the physical network diagram, (bottom left) a filtered version of the physical network showing a signal path of a selected signal, (right) a matrix view showing an overview of the logical network

and columns. Within this grid, RelEx not only marks the edges, but additionally encodes how many edges (signals) are exchanged between each pair of nodes. The number of signals exchanged is encoded by the size of a box that is drawn onto the matrix grid. With this box encoding important connections that exchange many signals visually pop out. This visual "pop out" effect is further supported by adding a black frame for the communication hotspots with more than 100 signals. We see that many signal boxes exist, that is, the logical network is very dense. In a node-link diagram, this characteristic would lead to extreme clutter making it almost impossible to perceive relevant information (the so-called hairball effect). Here, the matrix view offers a viable alternative. Note also, that the matrix is not symmetric as signals are usually unidirectional.

In addition, it supports tasks that might be harder to conduct with classical node-link representations. For instance, the nodes in the figure were ordered based on which hyperedge subsystem[1] they are connected to. These different subsystems are visually indicated as light blue background stripes in the matrix. Ordering the matrix in this way allows to better understand how much communication happens *within* a certain subsystem, and how much communications happens *across* different subsystems. This can be simply done by eyeballing the intersection between these bus stripes. The within communication is represented by the signal boxes that lie on the diagonal

---

[1]In automotive terminology, these are the bus systems that components are connected to, such as the CAN, MOST, or FlexRay bus. These domain-specific details are not relevant for the discussion here, and we hence simply refer to them as "subsystems."

intersections of stripes. For instance, the intersection rectangle at the top left, shown as blue highlight in Fig. 2.8, reveals a lot of communication within $subnet_1$. The orange highlight at the bottom right shows a different communication pattern. While there is also considerable communication within $subnet_2$, this system also receives (horizontally) and sends (vertically) many signals to other subsystems, as indicated by the dotted lines in Fig. 2.8. This example is meant to illustrate the importance of properly ordering matrix visualizations, so meaningful insights can be drawn from this visual representation.

2.4   Challenges and Opportunities

As the requirements change with origin, structure, content, representation, and interest, network visualization tasks abound and new challenges arise continuously. New approaches to network analysis and applications of network science in other domains inspire novel forms of network visualization.

Visualization tools therefore often combine tested and generic methods for layout with flexible means for attribute mapping and interactive exploration. Still, as visualization can be seen as the human lens to data, further research is needed to assess which visualization designs are understood by targeted groups of recipients. Display technologies, 3D printing, and augmented reality provide further opportunities to explore networks.

An important challenge for users of network visualization systems is not to fall for images of complexity and decorative beauty but to concentrate on the essential purpose of network visualization, namely to facilitate exploration and hypothesis formation as well as communication and the provision of evidence for conclusions.

References

Abello, J. & van Ham, F. (2004), Matrix zoom: A visual interface to semi-external graphs, *in* Ward & Munzner, pp. 183–190. https://doi.org/10.1109/INFVIS.2004.46

Bae, J. & Watson, B. (2011), 'Developing and evaluating quilts for the depiction of large layered graphs', *IEEE Trans. Vis. Comput. Graph.* **17**(12), 2268–2275. https://doi.org/10.1109/TVCG.2011.187

Behrisch, M., Bach, B., Riche, N. H., Schreck, T. & Fekete, J. (2016), 'Matrix reordering methods for table and network visualization', *Comput. Graph. Forum* **35**(3), 693–716. https://doi.org/10.1111/cgf.12935

Bertin, J. (1983), *Semiology of Graphics: Diagrams, Networks, Maps*, University of Wisconsin Press, Madison, WI.

Boettger, J., Brandes, U., Deussen, O. & Ziezold, H. (2008), 'Map warping for the annotation of metro maps', *Computer Graphics and Applications* **28**(5), 56–65.

Borgatti, S. P., Everett, M. G. & Johnson, J. C. (2018), *Analyzing Social Networks*, 2nd edn, Sage.

Brandes, U. & Pich, C. (2006), Eigensolver methods for progressive multidimensional scaling of large data, *in* M. Kaufmann & D. Wagner, eds, 'Graph Drawing, 14th International Symposium,

GD 2006, Karlsruhe, Germany, September 18-20, 2006. Revised Papers', Vol. 4372 of *Lecture Notes in Computer Science*, Springer, pp. 42–53. https://doi.org/10.1007/978-3-540-70904-6_6

Brandes, U. & Pich, C. (2008), An experimental study on distance-based graph drawing, *in* I. G. Tollis & M. Patrignani, eds, 'Graph Drawing, 16th International Symposium, GD 2008, Heraklion, Crete, Greece, September 21–24, 2008. Revised Papers', Vol. 5417 of *Lecture Notes in Computer Science*, Springer, pp. 218–229. https://doi.org/10.1007/978-3-642-00219-9_21

Brandes, U. (2016), Force-directed graph drawing, *in* M.-Y. Kao, ed., 'Encyclopedia of Algorithms', Springer, pp. 768–773. https://doi.org/10.1007/978-1-4939-2864-4_648

Brandes, U., Freeman, L. C. & Wagner, D. (2013), Social networks, *in* Tamassia, pp. 805–839.

Brandes, U., Indlekofer, N. & Mader, M. (2012), 'Visualization methods for longitudinal social networks and stochastic actor-oriented modeling', *Social Networks* **34**(3), 291–308. https://doi.org/10.1016/j.socnet.2011.06.002

Brandes, U., Lerner, J., Lubbers, M. J., McCarty, C. & Molina, J. L. (2008), Visual statistics for collections of clustered graphs, *in* 'IEEE Pacific Visualization Symposium', pp. 47–54.

Brandes, U. & Nick, B. (2011), 'Asymmetric relations in longitudinal social networks', *IEEE Transactions on Visualization and Computer Graphics* **17**(12), 2283–2290.

Díaz, J., Petit, J. & Serna, M. J. (2002), 'A survey of graph layout problems', *ACM Comput. Surv.* **34**(3), 313–356. http://doi.acm.org/10.1145/568522.568523

Dinkla, K., Westenberg, M. A. & van Wijk, J. J. (2012), 'Compressed adjacency matrices: Untangling gene regulatory networks', *IEEE Trans. Vis. Comput. Graph.* **18**(12), 2457–2466. https://doi.org/10.1109/TVCG.2012.208

Dwyer, T. (2009), 'Scalable, versatile and simple constrained graph layout', *Computer Graphics Forum* **28**(3), 991–998.

Elmqvist, N., Riche, Y., Riche, N. H. & Fekete, J. (2010), 'Melange: Space folding for visual exploration', *IEEE Trans. Vis. Comput. Graph.* **16**(3), 468–483. https://doi.org/10.1109/TVCG.2009.86

Gajer, P. & Kobourov, S. G. (2002), 'GRIP: graph drawing with intelligent placement', *J. Graph Algorithms Appl.* **6**(3), 203–224.

Gansner, E. R., Hu, Y., North, S. C. & Scheidegger, C. E. (2011), Multilevel agglomerative edge bundling for visualizing large graphs, *in* G. D. Battista, J. Fekete & H. Qu, eds, 'IEEE Pacific Visualization Symposium, PacificVis 2011, Hong Kong, China, 1–4 March, 2011', IEEE Computer Society, pp. 187–194. https://doi.org/10.1109/PACIFICVIS.2011.5742389

Gansner, E. R., Koren, Y. & North, S. C. (2004), Graph drawing by stress majorization, *in* Pach, pp. 239–250. https://doi.org/10.1007/978-3-540-31843-9_25

Ghoniem, M., Fekete, J. & Castagliola, P. (2004), A comparison of the readability of graphs using node-link and matrix-based representations, *in* Ward & Munzner (2004), pp. 17–24. https://doi.org/10.1109/INFVIS.2004.1

Grant, R. (2018), *Data Visualization: Charts, Maps, and Interactive Graphics*, CRC Press.

Hachul, S. & Jünger, M. (2004), Drawing large graphs with a potential-field-based multilevel algorithm, *in* Pach, pp. 285–295. https://doi.org/10.1007/978-3-540-31843-9_29

Hennig, M., Brandes, U., Pfeffer, J. & Mergel, I. (2012), *Studying Social Networks – A Guide to Empirical Research*, Campus.

Henry, N. & Fekete, J. (2007), Matlink: Enhanced matrix visualization for analyzing social networks, *in* M. C. C. Baranauskas, P. A. Palanque, J. Abascal & S. D. J. Barbosa, eds, 'Human-Computer Interaction—INTERACT 2007, 11th IFIP TC 13 International Conference, Rio de Janeiro, Brazil, September 10–14, 2007, Proceedings, Part II', Vol. 4663 of *Lecture Notes in Computer Science*, Springer, pp. 288–302. https://doi.org/10.1007/978-3-540-74800-7_24

Henry, N., Fekete, J. & McGuffin, M. J. (2007), 'Nodetrix: a hybrid visualization of social networks', *IEEE Trans. Vis. Comput. Graph.* **13**(6), 1302–1309. https://doi.org/10.1109/TVCG.2007.70582

Holten, D. (2006), 'Hierarchical edge bundles: Visualization of adjacency relations in hierarchical data', *IEEE Trans. Vis. Comput. Graph.* **12**(5), 741–748. https://doi.org/10.1109/TVCG.2006.147

Hu, Y. (2006), 'Efficient, high-quality force-directed graph drawing', *The Mathematica Journal* **10**(1), 37–71.

Kerren, A., Purchase, H. C. & Ward, M. O., eds (2014), *Multivariate Network Visualization—Dagstuhl Seminar #13201, Dagstuhl Castle, Germany, May 12–17, 2013, Revised Discussions*, Vol. 8380 of *Lecture Notes in Computer Science*, Springer. https://doi.org/10.1007/978-3-319-06793-3

Kivelä, M., Arenas, A., Barthelemy, M., Gleeson, J. P., Moreno, Y. & Porter, M. A. (2014), 'Multilayer networks', *J. Complex Networks* **2**(3), 203–271. https://doi.org/10.1093/comnet/cnu016

Kobourov, S. G. (2013), Force-directed drawing algorithms, *in* Tamassia, pp. 383–408.

Kruja, E., Marks, J., Blair, A. and Waters, R. C. (2001), A short note on the history of graph drawing, *in* P. Mutzel, M. Jünger & S. Leipert, eds, 'Graph Drawing, 9th International Symposium, GD 2001 Vienna, Austria, September 23–26, 2001, Revised Papers', Vol. 2265 of *Lecture Notes in Computer Science*, Springer, pp. 272–286. https://doi.org/10.1007/3-540-45848-4_22

Lietz, H., Wagner, C., Bleier, A. & Strohmaier, M. (2014), When politicians talk: Assessing online conversational practices of political parties on twitter, *in* 'Proceedings of the Eighth International Conference on Weblogs and Social Media, ICWSM 2014.', pp. 285–294. http://www.aaai.org/ocs/index.php/ICWSM/ICWSM14/paper/view/8069

Maaten, L. v. d. & Hinton, G. (2008), 'Visualizing data using t-SNE', *Journal of machine learning research* **9**(Nov), 2579–2605.

McGrath, C., Blythe, J. & Krackhardt, D. (1996), 'Seeing groups in graph layouts', *Connections* **19**(2), 22–29.

Munzner, T. (2014), *Visualization Analysis and Design*, AK Peters/CRC Press.

Newcomb, T. M. (1961), *The Acquaintance Process*, Holt, Rinehart, and Winston, New York, NY.

Nocaj, A., Ortmann, M. & Brandes, U. (2015), 'Untangling the hairballs of multi-centered, small-world online social media networks', *J. Graph Algorithms Appl.* **19**(2), 595–618. https://doi.org/10.7155/jgaa.00370

Ortmann, M., Klimenta, M. & Brandes, U. (2017), 'A sparse stress model', *J. Graph Algorithms Appl.* **21**(5), 791–821. https://doi.org/10.7155/jgaa.00440

Perry, B. L., Pescosolido, B. A. & Borgatti, S. P. (2018), *Egocentric Network Analysis*, Cambridge University Press.

Purchase, H. C. (2000), 'Effective information visualisation: a study of graph drawing aesthetics and algorithms', *Interacting with Computers* **13**(2), 147–162. https://doi.org/10.1016/S0953-5438(00)00032-1

Roberts, M. J. (2012), *Underground Maps Unravelled*, Self-published, Wivenhoe, UK.

Roweis, S. T. & Saul, L. K. (2000), 'Nonlinear dimensionality reduction by locally linear embedding', *Science* **290**(5500), 2323–2326.

Sedlmair, M., Frank, A., Munzner, T. & Butz, A. (2012), 'RelEx: Visualization for actively changing overlay network specifications', *IEEE transactions on visualization and computer graphics* **18**(12), 2729–2738.

Shen, Z. & Ma, K. (2007), Path visualization for adjacency matrices, *in* K. Museth, T. Möller & A. Ynnerman, eds, 'EuroVis07: Joint Eurographics—IEEE VGTC Symposium on Visualization, Norrköping, Sweden, 23–25 May 2007', Eurographics Association, pp. 83–90. https://doi.org/10.2312/VisSym/EuroVis07/083-090

Tamassia, R., ed. (2013), *Handbook on Graph Drawing and Visualization*, Chapman and Hall/CRC.

Tufte, E. R. (1983), *The Visual Display of Quantitative Information*, Graphics Press, Cheshire, CT.

van Garderen, M., Pampel, B., Nocaj, A. & Brandes, U. (2017), 'Minimum-displacement overlap removal for geo-referenced data visualization', *Computer Graphics Forum* **36**(3), 423–433. https://doi.org/10.1111/cgf.13199

von Landesberger, T., Kuijper, A., Schreck, T., Kohlhammer, J., van Wijk, J. J., Fekete, J. & Fellner, D. W. (2010), Visual analysis of large graphs, *in* H. Hauser & E. Reinhard, eds, 'Eurographics 2010—State of the Art Reports, Norrköping, Sweden, May 3–7, 2010', Eurographics Association, pp. 37–60. https://doi.org/10.2312/egst.20101061

Wang, Y., Wang, Y., Sun, Y., Zhu, L., Lu, K., Fu, C.-W., Sedlmair, M., Deussen, O. & Chen, B. (2018), 'Revisiting stress majorization as a unified framework for interactive constrained graph visualization', *IEEE transactions on visualization and computer graphics* **24**(1), 489–499.

Ware, C., Purchase, H. C., Colpoys, L. & McGill, M. (2002), 'Cognitive measurements of graph aesthetics', *Information Visualization* **1**(2), 103–110. https://doi.org/10.1057/palgrave.ivs.9500013

Wattenberg, M. (2006), Visual exploration of multivariate graphs, *in* 'Proceedings of the SIGCHI conference on Human Factors in computing systems', ACM, pp. 811–819.

Zinsmaier, M., Brandes, U., Deussen, O. & Strobelt, H. (2012), 'Interactive level-of-detail rendering of large graphs', *IEEE Trans. Vis. Comput. Graph.* **18**(12), 2486–2495. https://doi.org/10.1109/TVCG.2012.238

Chapter 3
A Statistician's View of Network Modeling

David R. Hunter

Abstract This introduction to statistical modeling of networks is aimed at an audience that possesses mathematical background at about the level of pre-calculus but that may not be familiar with what statisticians do. After illustrating the concept of statistical inference, the chapter discusses this concept in two main contexts where network data are analyzed: First, when a network is observed, and the aim is to learn about the process that may have formed it; and second, when the network itself is the object of scientific inquiry because it is unobserved.

3.1 Introduction

The development of statistical models for networks has a history spanning many decades, yet it is relatively recently that an explosion of interest in the study of networks within the statistical community has taken place. This brief glimpse of how statisticians view the modeling of networks is not intended to comprehensively survey the vast and rapidly growing literature on the subject, nor to outline the historical development of this work, but rather to introduce a reader who may have little previous exposure to the field of statistics to the way in which statisticians engage with the study of networks. While this discussion will not veer deeply into technical details, it does presume a comfort with mathematical ideas and notation at roughly the level of a pre-calculus course.

This chapter discusses the statistical paradigm itself in Sect. 3.2 and introduces some notation in Sect. 3.3. Then, Sects. 3.4 and 3.5 present two fundamentally different network modeling scenarios with which statisticians concern themselves. This chapter is by no means meant to be an exhaustive survey of statistical network modeling, and the fact that Sect. 3.4 is substantially longer than Sect. 3.5 merely reflects my own particular areas of expertise rather than a bias in the statistical literature generally.

D. R. Hunter (✉)
Department of Statistics, Penn State University, State College, USA
e-mail: dhunter@stat.psu.edu

© Springer Nature Switzerland AG 2019
F. Biagini et al. (eds.), *Network Science,* https://doi.org/10.1007/978-3-030-26814-5_3

3.2 Statistical Inference

Statistics is sometimes understood by the non-specialist as a collection of techniques for summarizing data, or even as summaries of data themselves: I remember a call I once received while working the phones as a graduate student in the University of Michigan's statistical consulting center in which the caller, after verifying that he had in fact reached a statistician, asked whether I could tell him the average annual rainfall in the Amazon rainforest. Of course, summaries of large amounts of data, whether visual or numeric, are very useful—yet perhaps unsurprisingly, such summaries are merely means used by statisticians to achieve more nuanced ends, rather than the ends themselves.

In its most basic formulation, the science of statistics consists of methods for learning about a population from a sample. Sometimes, the population may be a well defined, if difficult to enumerate, group of individuals—for instance, every citizen of France over the age of eighteen. Other times, the population is conceptual, such as the theoretical set of all possible flips of a fair coin. One interesting aspect of the statistical study of networks, as we shall see, is that the definition of the population of interest is not always immediately obvious from the context of a particular application.

Consider the game of craps, which involves rolling a pair of (presumably fair) six-sided dice. With each roll, only the total number of dots shown is recorded, and the game ends, after one or more rolls, with the result being either a win or a loss for the player.[1] It turns out that the probability of a win in a game of craps is exactly 244/495, which means the probability of a loss is 251/495.

Knowing that the probability, call it p, of winning a game of craps is 244/495, we can invoke basic probability concepts to describe, say, the possible outcomes if we play 100 consecutive games of craps. For instance, we might let the variable X denote the total number of wins in 100 consecutive games of craps. In this case, we call X a random variable because it assigns a real number to each possible outcome of our 100-game experiment, and probability allows us to use the value of p to describe the likelihood of the various possible values X could take. Notice in this simple example that we tacitly accept, based on our physical knowledge of what happens when we roll dice, that none of our 100 games of craps has any influence on any of the other games. In probability parlance, we say that the games are therefore *independent*, and the notion of probabilistic independence will come up multiple times in this chapter.

Statistics, on the other hand, is often described as "probability in reverse" because in a statistical situation, we use observed data X to learn what we can about the population parameter p. That is, we conceive of an infinite population consisting of all possible games of craps, from which we select a representative and independent sample of size 100 simply by playing the game 100 times, then consider what our observed proportion of wins tells us about the proportion, p, that exists in the

---

[1]More precisely: If the first roll is a 2, 3, or 12, the game ends immediately in a loss; if the first roll is a 7 or 11, the game ends immediately in a win. In any other case, the result of the first roll is called the "point," and then the game changes starting at the second roll onward, with the "point" resulting in a win, a 7 resulting in a loss, and any other roll resulting in merely another roll.

population. In the case of craps, this may seem uninteresting, since we can already calculate p from first principles and an understanding of the properties of rolls of a fair six-sided die. Yet in many situations, the parameters that summarize some property of interest about a population are unknown, and the job of statistics is to learn about them using data derived from a sample. This process of learning about a population from a sample is called statistical inference, and it is the development of methods for statistical inference—rather than amassing facts about Amazon rainfall—that is the purview of academic statisticians.

3.3   Terminology and Notation

In this chapter, we will use "network" and "graph" roughly interchangeably, and each term may change its meaning slightly depending on the context. For instance, we will not use the term "graph" merely in the strict mathematical sense of a set of pairs of elements from a given vertex set. Sometimes, a graph might include not only a set of ordered or unordered pairs of vertices, but also one or more variables measured on each vertex. This is the sense in which "graph" is used, for example, by (Handcock et al. 2017) in the software package for R (R Core Team 2018) that coined the phrase "exponential-family random graph model." By contrast, (Fellows and Handcock 2013) defined an "exponential-family random network model," or ERNM. Yet for our purposes, there is really no difference between graph and network; the differences between ERGM and ERNM have to do with the respective forms of the models rather than the object of the modeling. In addition, the terms "vertex" and "node" will be used interchangeably in this article.

As shown in Fig. 3.1, we will often use Y to denote a matrix giving the essential structure of a network, or graph, with Y_{ij} giving the state of the connection between nodes i and j. In the case that the network is directed, the values of Y_{ij} and Y_{ji} might be different; otherwise, we always have $Y_{ij} = Y_{ji}$ and so we may restrict our

Fig. 3.1 A five-node directed network consisting of the seven ordered pairs (1, 4), (2, 1), (2, 3), (3, 1), (3, 2), (3, 4), and (5, 4) may be expressed as a figure (left) or as an adjacency matrix (right)

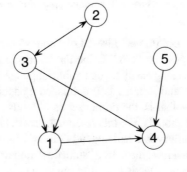

$$Y = \begin{pmatrix} 0 & 0 & 0 & 1 & 0 \\ 1 & 0 & 1 & 0 & 0 \\ 1 & 1 & 0 & 1 & 0 \\ 0 & 0 & 0 & 0 & 0 \\ 0 & 0 & 0 & 1 & 0 \end{pmatrix}$$

attention to the upper triangular entries Y_{ij} where $i \leq j$. In addition, we may or may not omit the possibility of self-edges, i.e., cases in which $Y_{ii} \neq 0$. By default, we will assume that Y_{ij} is equal to 0 or 1, so that the network is binary; yet in some cases, we will consider weighted edges in which Y_{ij} takes values other than 0 and 1.

3.4 Inference About Network Model Parameters from Network Data

As (Kolaczyk 2017) eloquently states in a recent monograph about topics at the frontier of statistics and network analysis, we concern ourselves in this chapter with "data either *of* or *from* a system conceptualized as a network." Within this section, the data are not merely from the system but rather the network itself. That is, we observe directly either the values of all Y_{ij}—except perhaps for some missing values—or at least some function of the Y_{ij}. In addition, we may have additional measurements on the nodes, or edges, or both. This section defines the idea of a statistical model and discusses some of the many ways in which statisticians build models for data of this type.

3.4.1 The Erdős-Rényi-Gilbert Model

Perhaps the simplest statistical model for a network is the one that is described by (Gilbert 1959) as follows:

> Let N points, numbered 1, 2, ..., N, be given. There are $N(N-1)/2$ lines which can be drawn joining pairs of these points. Choosing a subset of these lines to draw, one obtains a graph; there are $2^{N(N-1)/2}$ possible graphs in total. Pick one of these graphs by the following random process. For all pairs of points make random choices, independent of each other, whether or not to join the points of the pair by a line. Let the common probability of joining be p. Equivalently, one may erase lines, with common probability $q = 1 - p$ from the complete graph.

Aside from the use of "points" instead of "nodes" or "vertices" and the use of "lines" instead of "edges" or "ties," Gilbert's description assigns a probability to every possible network using easily recognizable language. Contemporaneously with Gilbert's paper, (Erdős and Rényi 1959) published a paper that studied some of the asymptotic properties—that is, the properties as N tends to ∞—of the same model as well as other similar models. For this reason, the model is often called the "Erdős-Rényi model." Here, however, I will append Gilbert's name to the title.

First, let us discuss what we mean by a "statistical model": In Gilbert's explanation, notice that each choice of p leads to a different way to assign probabilities to every one of the $2^{N(N-1)/2}$ possible networks. Each such assignment is called a probability *distribution* on the set of possible networks, and a statistical model is nothing more than a set of probability distributions. Associated with a model, we often have a *parameterization*, which is a way to associate each probability distribution with a real number or vector; in this case, the parameter is p and the association is implicit in Gilbert's description. When every distribution in a model is associated with a *unique* parameter value or vector, we say that the parameterization is *identifiable*, which is a necessary feature of any situation that will allow for statistical inference. (If a particular data-generating distribution is associated with more than one parameter, then no amount of information from data will enable one to distinguish among certain parameter values.)

In Gilbert's description, just as in the craps example, if we know p then we may describe the probabilities associated with any possible value of the (yet-to-be-observed) network Y. But statistics is probability in reverse: We observe Y and the goal is to learn about the value p that gave rise to it. Mathematically, all of the information we may obtain about p from observing Y is contained in a single value measured on Y, namely the number of lines, or edges, drawn. Let us label this statistic $s(Y)$; using the convention that Y_{ij} is equal to 0 or 1 for all $i < j$, we could write

$$s(Y) = \sum_{i=1}^{n-1} \sum_{j=i+1}^{n} Y_{ij}.$$

Because $s(Y)$ conveys all the information about p contained in Y itself, we say that $s(Y)$ is a *sufficient statistic* for this model.

Thinking of Y as a random, yet-to-be-observed network whose specific configuration will be determined according to the method described above by Gilbert, let us denote by the lowercase letter y a particular configuration that could arise. Then, $s(y)$ is the number of edges in y, and the probabilities associated with all network configurations y having identical $s(y)$ values are the same: Each is equal to $p^{s(y)} q^{N(N-1)/2 - s(y)}$, where we define q to be equal to $1 - p$ to simplify notation. Rearranging slightly, we may write this statement as

$$\Pr_p(Y = y) = \left(\frac{p}{q}\right)^{s(y)} q^{N(N-1)/2},$$

where $\Pr_p(A)$ means the probability of an event A assuming that the parameter takes value p. For mathematical convenience, let us define $\theta = \log(p/q)$, the natural logarithm of the odds p/q that any given edge exists. We may then write

$$\Pr_\theta(Y = y) = \exp\{\theta s(y)\} q^{N(N-1)/2} = \frac{\exp\{\theta s(y)\}}{\kappa(\theta)}, \tag{3.1}$$

where we define $\kappa(\theta)$ as $1/q^{N(N-1)/2}$, recalling that $q = 1 - p$ is a function of θ. Importantly, $\kappa(\theta)$ does not depend on the sufficient statistic $s(y)$. The model whose parameterization is presented in (3.1) is called an exponential-family model because it expresses a probability distribution as proportional to the exponentiation of a parameter θ times a statistic s. In general, an exponential-family model may take θ and s to be vectors, and in this case, it is their dot product that is exponentiated.

In the case of the Erdős-Rényi-Gilbert model as expressed in (3.1), where $s(y)$ is the number of edges in y, despite all of the development in this section we are merely in the same situation described in the craps example: We observe a certain number $s(y)$ of successes in an experiment consisting of $N(N-1)/2$ trials, and we are trying to learn about the parameter p—or, equivalently, θ—representing the probability of success in each trial. Indeed, just as in the craps example, an intuitively appealing estimator of p is the proportion of observed successes. If we let y_{obs} denote the observed network, then the proportion of observed successes is $2s(y_{obs})/N(N-1)$.

For example, in Fig. 3.2 we see a modified version of the network depicted in Fig. 3.1, where the relationships between pairs of nodes are undirected instead of directed. In this example, $N = 5$ since there are 5 nodes and $s(y_{obs}) = 6$ since there are 6 undirected edges observed in our network out of the $(5 \times 4)/2$, or 10, possible such edges. This gives a proportion of observed successes equal to 0.6. Another way to state this is that our estimate of p, sometimes denoted by $\hat{p}$ where the "hat" above the parameter p shows that $\hat{p}$ is a data-based estimate (pronounced "p-hat") of the parameter p. Furthermore, since θ is easily expressible as a function of p, namely $\theta = \log(p/[1 - p])$, we may write the estimate of θ as $\log(0.6/0.4)$ in our example, which equals 0.405 (remember, the logarithm here is the natural logarithm). In a general problem involving the assumption that an Erdős-Rényi-Gilbert model gave rise to an observed undirected network, we may express the estimate of θ as

$$\hat{\theta} = \log \left(\frac{2s(y_{obs})}{N(N - 1) - 2s(y_{obs})} \right). \tag{3.2}$$

In statistical inference, we are rarely content with merely an estimator of the parameter of interest; we also want a sense of how precise that estimator is. After all, any estimator is based on data, which are variable, so the estimator itself is also variable. Often, the precision of an estimator is expressed as an estimate of

Fig. 3.2 A five-node undirected network consisting of the six unordered pairs $\{1, 2\}$, $\{1, 3\}$, $\{1, 4\}$, $\{2, 3\}$, $\{3, 4\}$, and $\{4, 5\}$ may be expressed as a figure (left) or as an adjacency matrix (right) in which only the above-diagonal entries are necessary since the matrix is symmetric

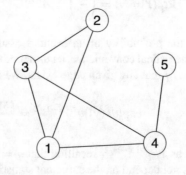

$$Y = \begin{pmatrix} 0 & 1 & 1 & 1 & 0 \\ 1 & 0 & 1 & 0 & 0 \\ 1 & 1 & 0 & 1 & 0 \\ 1 & 0 & 1 & 0 & 1 \\ 0 & 0 & 0 & 1 & 0 \end{pmatrix}$$

the standard deviation of the estimator itself, called the standard error. To describe methods for estimating precision of an estimator would take the current article too far afield. An interested reader could search for material on maximum likelihood estimators—since $\hat{\theta}$ in our example is an example of an MLE—and how the precision of MLEs is estimated.

On the other hand, one statistical question we will address, which is much more complicated in the network context than in most traditional data collection frameworks, is this: What is the sample size? Is it the number of nodes, N, or the number of possible edges, $N(N-1)/2$, or the number of networks that we observe, which is often just 1? The answer to this question is important, since the aforementioned method of maximum likelihood estimation is largely justified based on theoretical results that describe the behavior of these estimates as the sample size tends to ∞, a subject often referred to as large-sample, or asymptotic, statistics. In the case of the Erdős-Rényi-Gilbert model, our estimator behaves as though the sample size is the number of possible edges. In fact, this is true for many types of network models in which the status of each node pair—whether or not it has an edge—is independent of all other node pairs. However, in general answering this sample size question is tricky, as we shall see in the next section.

3.4.2 A Generalization of Erdős-Rényi-Gilbert

It is not hard to extend the Erdős-Rényi-Gilbert model in multiple directions. One such generalization is alluded to in the previous section: An exponential-family model with d-dimensional parameter vector θ and d-dimensional sufficient statistic $s(y)$ takes the form

$$\Pr_\theta(Y = y) = \frac{\exp\{\theta^\top s(y)\}}{\kappa(\theta)}, \tag{3.3}$$

where $\theta^\top s(y) = \sum_i \theta_i s_i(y)$ is the usual dot product of the vectors θ and $s(y)$, and (3.3) is usually called an exponential-family random graph model (ERGM) in the modern statistical literature. Special cases of (3.3) have a decades-long history, and the general form of this model was originally called a p-star model by (Wasserman and Pattison 1996) before the abbreviation "ERGM" became popular. For readers interested in this history, several surveys exist, such as the article by (Goldenberg et al. 2009) and the book-length treatment of network models by (Kolaczyk 2009). Since the dimension d of $s(y)$ can take any value we wish, it is possible to incorporate into the model of (3.3) any set of sufficient statistics that we feel are relevant to the probability of a particular network configuration.

There are myriad possibilities. For instance, suppose that we measure a variable for every node that places each node into a category, such as gender. If one of the d entries of $s(y)$ is the number of edges that connect two nodes that belong to the same category, then the θ associated with this statistic describes the propensity toward homophily on this variable—that is, the tendency for like individuals to associate with one another—in the observed network. As another example, suppose that another of the entries of $s(y)$ is the number of two-paths contained in y, that is, the number of pairs of (undirected) edges emanating from the same node. This situation is depicted and explained in Fig. 3.3. A lengthy list of entries a researcher might consider including in the $s(y)$ vector is presented by (Morris et al. 2008).

In the special case of the model of (3.1), we saw that there is a simple expression for both the function $\kappa(\theta)$ and the estimator $\hat{\theta}$. This estimator has the property that it maximizes the value of $\mathrm{Pr}_\theta(Y = y_{\mathrm{obs}})$ as a function of θ; it is thus, by definition, a maximum likelihood estimator. Furthermore, for an exponential-family model, the maximum likelihood estimator has the intuitively pleasing property that it is the unique parameter value for which the expected value of the random vector $s(Y)$ is exactly equal to the observed vector $s(y_{\mathrm{obs}})$. Unfortunately, however, (3.1) with its corresponding closed-form expression for $\hat{\theta}$ in (3.2) is an unusual case; in general, neither $\kappa(\theta)$ nor $\hat{\theta}$ may be expected to have a simple form. Indeed, $\hat{\theta}$ may be exceedingly difficult to calculate in practice. We will not describe the difficulties here; interested readers may refer to (Hummel et al. 2012) and the references therein for a technical treatment of the details.

Statistical modeling of networks using ERGMs highlights a vexing issue for network analysis generally in the statistical context, namely the question of what exactly is the population from which we are sampling. Since individuals are the sampled units in a statistical population, and since individuals often play the role of nodes in a network context, it seems logical that we should view the population of individuals as providing the basis for our sample, and somehow we manage to observe all of the relevant ties between pairs of individuals that happen to be selected for

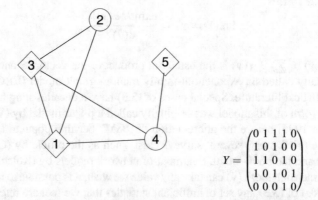

Fig. 3.3 Here, each node is observed to belong to one of two categories, depicted as circles and squares. Suppose that in (3.3) the $s(y)$ vector has three entries: $s_1(y)$ is the number of edges in y, $s_2(y)$ is the number of edges connecting matching nodes in y, and $s_3(y)$ is the number of two-paths in y. From the figure, we see that $s_1(y_{\mathrm{obs}}) = 6$ and $s_2(y_{\mathrm{obs}}) = 1$ and $s_3(y_{\mathrm{obs}}) = 10$. In this particular 3-term ERGM, unlike in the simple Erdős-Rényi-Gilbert example, there are no easily derivable closed-form expressions for the maximum likelihood estimators $\hat{\theta}_1$, $\hat{\theta}_2$, and $\hat{\theta}_3$

our sample. However, this formulation is problematic for several reasons. First, it may not happen that network data arise in this way; instead, the particular nodes we observe are in the sample precisely because they are part of a network of interest, so these nodes might not be considered a representative sample from the population of individuals (for instance, because our network framework tends to undersample from the population of isolated nodes). Second, the Erdős-Rényi-Gilbert example shows that even in the simplest of all network models, the sample size is clearly $N(N - 1)/2$, not N. In other words, when our model assumes that the edges form independently of one another, we obtain statistical power as though our population consisted somehow of *pairs* of individuals, rather than the individuals themselves— yet we clearly cannot sample an arbitrary set of $N(N - 1)/2$ such pairs from the population, as we are constrained to exactly those pairs that arise from the N individuals that comprise our observed network.

Finally, if our population is a set containing the nodes in our sampled network, the very basis for statistical inference suggests that we should be able to learn about properties of the whole population from the sample we have selected. In the ERGM framework, the "properties" are the parameter values: We assume that if our model is valid globally, then there are true parameters and that our sample helps us make inference about these values. However, mathematically this is provably *not* what happens in all cases; (Shalizi and Rinaldo 2013) outline some of these limitations in a highly technical article.

One way around these issues is to redefine the population. Instead of sampling from a population of nodes, we imagine the nodes in the network to be fixed, and we sample a single observation from the theoretical population of all possible networks on those nodes. Of course, in general not all networks on a given set of nodes are equally likely, but it is precisely the purpose of the statistical model to define a family of probability distributions on these networks. One possible distribution does give equal probability to each possible network, namely the one in which $\theta = 0$ in (3.1) or (3.3), but in general each θ parameter value gives either more or less probability mass, depending on its sign, to networks for which the corresponding $s(y)$ statistic is higher assuming all other statistics remain constant. One implication of this framework is that our sample size is the number of networks we observe, which is generally just one! (We could conceivably observe multiple independent networks on the same set of nodes, but this is rare.) Since many of the nice properties of the statistical estimators we use are justified based on their asymptotic, or large-sample, properties, this fact should give us pause. Yet it is also the case that a single network observation can contain a great deal of information about the parameter values—keep in mind the Erdős-Rényi-Gilbert example in which a single network contains the information of a sample of size $N(N - 1)/2$—so our faith in techniques such as maximum likelihood estimation is not misplaced.

3.4.3 Network Models Involving Latent Variables

Quite often, data are only partially observed. This is frequently due to actual missing data, that is, cases in which we try but fail to obtain measurements of a particular variable. Yet sometimes unobserved, or latent, variables arise by design: For instance, it is quite common in statistical analysis of data on educational testing to assume that each question on an examination has some abstractly defined notion of "difficulty" associated with it, and we are only able to observe this difficulty indirectly through the scores of numerous test-takers who are each exposed to some subset of questions. In such cases, statisticians have developed methods for dealing with these latent variables, whether or not the estimation of these variables is of particular interest.

One of the most common ways in which latent variables arise is in the context of a statistical mixture model. Mixture models help us solve the so-called unsupervised clustering problem of multivariate analysis that has myriad non-statistical solutions as well: We observe data $X_1, \ldots, X_N$ that are often vector-valued (though they may be scalar-valued), one observation for each of N individuals. We assume that each individual is a member of one of K categories, but we do not observe the category memberships. The unsupervised clustering problem is the problem of trying to discover the category memberships using only the observed data.

In the context of network models, the classic example of a mixture model is the so-called stochastic block model (Snijders und Nowicki 1997), in which every node is assumed to belong to exactly one of K categories. Conditional on the category memberships, we assume that every potential edge arises independently of the others with a probability that is solely determined by the memberships of its two nodes. In its most general form, this model defines a separate probability parameter governing the formation of ties for each of the $K(K + 1)/2$ possible pairs of node-pair category memberships. Sometimes, it is assumed that edges within the same category are more likely than edges between categories; that is, $p_{ii} > p_{ij}$ whenever $j \neq i$, where p_{ij} denotes the probability of an edge between two nodes in categories i and j.

Generally, mixture models are easily amenable to statistical analysis. For instance, estimation of the parameters in a mixture model, including a K-length vector for each individual giving the respective probabilities of membership of that individual in each of the K groups, is often accomplished by maximum likelihood estimation using a computational technique called the EM algorithm. Yet the stochastic block model, as is the case with so many statistical network models, foils standard approaches to obtain maximum likelihood estimates. Even the EM algorithm in its usual form turns out to be intractable for stochastic block models, so we must rely on approximation methods for statistical estimation. We do not present any of these methods here, but interested readers will find a relevant discussion in the survey article of (Hunter et al. 2012).

One way to characterize the stochastic block model is that the connectivity of each node depends on a particular characteristic for that node, where the characteristic can take only a fixed number of values, namely the number of categories. If, instead of a discrete value, each node is associated with a continuous latent value that determines its connectivity, then we have essentially the idea of the random-effects model. In its most simplistic form, we might assume that the probability of an edge between nodes i and j exists with probability whose log-odds is equal to $\alpha_i + \alpha_j$, where all α_i values are realizations from a continuous distribution whose parameters are the object of our estimation. An extension of this basic idea to the case of directed networks is discussed by (van Duijn et al. 2004).

In a more advanced version of this idea, suppose that each node is associated with a point in k-dimensional space. This latent variable may be thought of as some kind of position in social space that has some bearing on the network edges that are formed between nodes beyond what we are able to explain using variables that we can observe. Our statistical model then includes some way to incorporate "distance" between two nodes in this latent space into the probability of an edge between them. This "distance" might be Euclidean, as studied by (Handcock et al. 2007), or cosine (dis)similarity, as in (Hoff et al. 2002).

Finally, we return to the idea of the first paragraph in this subsection, namely that sometimes data are simply missing even when we expect that they should be there. An adequate treatment of this subject, which is an area of statistical research unto itself, is beyond the scope of this brief chapter. Yet it is at least necessary to point out one of the most important aspects of any missing data analysis, which is that an unobserved Y_{ij} value is *not* equivalent to the case $Y_{ij} = 0$, or the observed absence of an edge. One convenient method for correctly dealing with missing data is presented by (Handcock and Gile 2010), who introduce a second network, D, with entries defined so that $D_{ij} = 1$ if Y_{ij} is observed, and $D_{ij} = 0$ otherwise. They even introduce a model for the D network, showing that the estimation of parameters for the Y model is possible using techniques that are straightforward in principle as long as the D network satisfies a technical condition called *missing at random*, or MAR: In fact, MAR is not quite what it sounds like; the entries in D are allowed to depend on the values of Y, both observed and unobserved. (If D is independent of Y, then we say that the data are *missing completely at random*, or MCAR, which is a stronger condition than MAR.) What MAR means is that all of the information in Y about the pattern of missing data is contained in the observed set of Y_{ij} values. Letting Y_{obs} and Y_{mis} denote the observed and missing portions of the Y network, we may equivalently say that MAR means that conditional on Y_{obs}, D is independent of Y_{mis}, or

$$P(D = d \mid Y_{\text{obs}}, Y_{\text{mis}}) = P(D = d \mid Y_{\text{obs}})$$

for any possible value d that D could take. Though the MAR condition can be difficult to check in practice, it is realistic in many applications to assume that when data are missing, they are missing for reasons unrelated to the values that would have been observed.

3.4.4 Models for Time-Varying Networks

It is often the case that networks are observed over time, and the patterns of connections between nodes change with time. One way a time-varying network dataset could arise is when the same network is observed at multiple discrete time points. Yet as noted in Subsect. 3.4.2, it is relatively rare to have data consisting of multiple observations of a network on the same set of nodes. When such data are observed, of course it is possible to apply any of the modeling methods described previously to each time point separately; but to do so ignores the temporal structure inherent in the dataset. As a simplistic approach to accounting for time evolution of the network, one might augment the ERGM of (3.3) so that the left-hand side of that equation is the probability that the network takes configuration y at time $t + 1$ given that at the previous time step t it takes the configuration y'. The statistics upon which this conditional probability might depend will presumably be measured on both y and y', and (3.3) becomes

$$\Pr_\theta(Y^{(t+1)} = y \mid Y^{(t)} = y') = \frac{\exp\{\theta^\top s(y, y')\}}{\kappa(\theta)}. \tag{3.4}$$

The modeling framework implied by (3.4) is essentially the approach taken by (Hanneke et al. 2010) and, in a slightly more complicated form, (Krivitsky and Handcock 2014), and it is sometimes referred to as a *Markovian* model, which means that random behavior of the system at time $t + 1$ depends on the past only through its state at t (and not on any other past behavior).

As an alternative approach to network data observed at multiple time points, sometimes called *longitudinal* network data, the stochastic actor-based models developed by Tom Snijders and several of his collaborators (see, e.g., (Snijders 2017)) model the individual behavior of each node in the network according to its propensity to form ties with each of the other nodes, which is expressed as some function of statistics, selected by the researcher, on the pair of nodes involved in the potential tie in question and/or the network configuration immediately surrounding those nodes. This actor-based model is assumed to proceed in continuous time, even if the data are only observed at discrete time points. In particular, for any possible time t, the time increment until the next event has a random distribution known as the *exponential distribution* with interesting mathematical properties that aid considerably in the estimation of these models. Among these properties are the fact that taking the minimum of a set of independent exponential random variables produces a random variable whose distribution is once again exponential, and also the fact that an exponentially distributed waiting time is *memoryless* in the sense that at all times, the distribution of time remaining until an event occurs is the same regardless of how long one has already waited. (If this latter property sounds paradoxical, consider the simple experiment of flipping a coin until you see the first occurrence of "heads": The random properties of the number of flips required to attain this goal are exactly the same whether we have already flipped the coin twice, or ten times, or not at all.

An exponential distribution is merely the continuous-time analogue of this discrete coin-flipping experiment.) Snijders' well-established stochastic actor-based model, which has a history spanning more than a decade, is implemented in the publicly available RSiena package (Ripley et al. 2017) for R (R Core Team 2018).

While it is rare to observe a full network at multiple discrete time points, it is increasingly common in the age of online social networks to observe data in which each edge has a time stamp indicating *when* that edge was initiated or occurred. Indeed, even nodes themselves may enter and/or exit the observable network at known time points. In this context, we may wish to model the instantaneous frequency or rate of events, as a function of whatever measured covariates we deem to be relevant, as this rate changes over time. To this end, researchers in statistics have recently begun to apply a well-established class of models, referred to in various contexts as survival models or reliability models or counting process models, to network datasets. I will use the term "counting process" here. The basic mathematical idea is explained by (Butts 2008), and to cite just one example of an application, (Perry and Wolfe 2013) use counting processes to analyze a network of time-stamped email contacts among a set of individuals in a company.

The central object of interest of counting process models is the rate of edge formation, which will be denoted $\lambda_{ij}(t)$ for the edge from i to j at time t. This rate may be interpreted as follows: For small values of δ, the probability that an edge will occur between nodes i and j in the time from t to $t + \delta$ is roughly $\delta \times \lambda_{ij}(t)$. The myriad methods within the field of survival analysis allow researchers to express $\lambda_{ij}(t)$ as a function of certain covariates in much the same way that an ERGM or an actor-based model allows one to express the probability of an edge between two nodes as a function of certain covariates. These functions, as well as the covariates themselves, may or may not change over time.

3.4.5 A Few Words About Software

Any attempt to provide an extensive catalog of existing software for implementing the statistical models presented here is doomed to be incomplete; there are simply too many software packages available, and their number is ever-expanding as new models and methods are developed. However, since this chapter is explicitly about a statistician's viewpoint, there are a couple potentially useful points to make here.

First, there is a large amount of software devoted to calculating various statistical summaries of networks and producing graphical representations of networks. None of this software is directly relevant to this chapter. As stated in Sect. 3.2, this chapter seeks to distinguish between "statistics" used as a plural noun, that is, summaries of data, and "statistics" as a scientific discipline, by which we usually mean statistical inference. As for graphical representations, the algorithms used to render, say, an adjacency matrix as a two-dimensional "dots and lines" figure by placing the dots in such a way as to achieve a pleasing-looking diagram do not involve statistical models and are therefore unrelated to the topic of this chapter.

It is safe to say that the majority, though certainly not all, of the software written these days that implements statistical inference is written for the R environment (R Core Team 2018). A notable exception is the PNet software (Wang et al. 2006), produced at the University of Melbourne, which implements estimation methods for ERGMs and other statistical network models (http://www.melnet.org.au/pnet/). The statnet suite of packages for R (Handcock et al. 2008) implements myriad different statistical network models, including ERGMs, certain types of latent variable models, and certain types of time-varying network models; information on these packages is available at https://statnet.org/trac/wiki/StatnetComponents. As mentioned in Subsect. 3.4.4, the RSiena package for R (Ripley et al. 2017) implements a particularly well-known class of time-varying network models, also known as longitudinal network data models, among other statistical models for social networks; additional information is at https://www.stats.ox.ac.uk/~snijders/siena/. One helpful feature of an open-source software environment such as R is that it allows for the possibility of independently developed yet interoperable packages. For instance, there are some R packages, such as the Bergm package (Caimo and Friel 2014) that implements a Bayesian estimation framework for ERGMs, that are built to depend on one or more of the statnet packages even though they are not themselves officially part of the statnet suite of packages. Finally, multiple stand-alone R packages implement certain specific statistical models for networks; to take just one example, the stochastic block models of Subsect. 3.4.3 and their various extensions are implemented in the R packages blockmodels (INRA and Leger 2015) and blockmodeling (Žiberna 2018).

3.5   Inference About Networks from Multivariate Data on the Nodes

The notion that statistical analysis of networks involves "data either *of* or *from* a system conceptualized as a network" (Kolaczyk 2017) does not necessarily imply that the data are observations of the network itself. For instance, (Groendyke et al. 2012) consider a dataset consisting merely of dates of onset of and recovery from measles for a group of children in a German town in 1861, estimating parameters for a statistical model for a network of contacts among the children that could have given rise to the outbreak.

Yet this section considers a different type of situation, in which the data are vector-valued observations made on a set of n entities. If we let d denote the dimension of the vectors, then d is the number of nodes (here, the sample size n is not relevant to the network structure). Thus, the data consist of vectors $X_1, \ldots, X_n$, each with d dimensions, and each dimension is associated with exactly one node. The edges in this network are defined by the notion of conditional dependence: An edge Y_{ij} exists if and only if the corresponding node values are conditionally dependent given the variables associated with all other nodes. This means that the value measured on dimension i contains information about the value measured on dimension j even if all information about j contained in all of the other dimensions is taken into account. The goal of this type of work, then, is to try to determine the structure of the network Y based solely on the d-dimensional observations from a sample of size n.

3.5.1 Learning Networks via Precision Matrices

Consider the example described in (Sachs et al. 2005), which studied a dataset involving multivariate measurements on more than eleven thousand human immune system cells. In particular, a multiparameter flow cytometer simultaneously recorded levels of eleven phosphoproteins and phospholipids, so that in this problem $d = 11$—that is, our network consists of eleven nodes, with each node corresponding to a particular protein or lipid.

Various cells had been perturbed with various molecular interventions, and the overall goal was to look for relationships between pairs of the 11 proteins as expressed through their measurements in this large group of cells. In particular, we consider two proteins (nodes) to be related if their expression levels are conditionally dependent, which means that knowledge of the value of one of the levels provides some information about the value of the other, even if we take into account all the information contained in the other $d-2$ expression levels. We want to learn about the entire network of these two-variable relationships among the 11 nodes, since the existence of such relationships could reveal scientifically important cellular signaling mechanisms.

On the right side of Fig. 3.4, we see the directed network that was reported by (Sachs et al. 2005). In this network, each node is labeled with the name of a particular protein or lipid, and an arrow from one to another indicates that the expression level of the first appears to affect the expression level of the second. This type of directed network, which is frequently constrained so that it cannot include any set of edges forming a circular loop—which makes it "acyclic"—is sometimes called a Bayesian network. Such networks are described in more detail in Chap. 6 of this book.

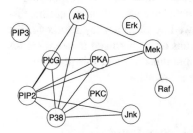 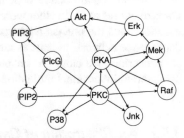

Fig. 3.4 On the right is the Bayesian network estimated by (Sachs et al. 2005). On the left is one of the undirected conditional dependence networks estimated by (Friedman et al. 2008)

It is important to understand that directionality of links in a Bayesian network cannot in general be established merely using the information contained in the observations themselves. Indeed, in (Sachs et al. 2005), the $n \times d$ dataset was combined with additional biological information about how the cells had been perturbed and what these perturbations implied regarding the possible direction of the arrow, if it exists, between any given pair of nodes.

In this chapter, we consider only the data themselves, which means that directionality information is incomplete at best and so the goal is to learn the network of undirected interrelationships among various pairs of nodes. This is exactly the approach taken by (Friedman et al. 2008), who considered a subset of $n = 7466$ of the original 11,672 eleven-dimensional observations. One of the undirected networks inferred by those authors is depicted on the left side of Fig. 3.4.

Much of the statistical literature on estimating the conditional dependence network Y based on d-dimensional multivariate observations X is based on the elegant mathematical theory of the multivariate normal distribution. A comprehensive discussion of the multivariate normal is far beyond the scope of this article, but in brief, we will say that the random vector X has a multivariate normal distribution with mean vector μ and covariance matrix Σ if the probability density function of X may be expressed, for all possible d-dimensional values x, as

$$f(x) = \frac{1}{(2\pi)^{d/2}\sqrt{\det \Sigma}} \exp\left\{-\frac{1}{2}(x-\mu)^\top \Sigma^{-1}(x-\mu)\right\}. \qquad (3.5)$$

In (3.5), μ must be d-dimensional and Σ must be a symmetric $d \times d$ matrix that is strictly positive definite (that is, all of the eigenvalues of Σ must be positive), which implies among other things that Σ has a well-defined matrix inverse, Σ^{-1}. This inverse matrix is quite important, and we shall write $\Omega = \Sigma^{-1}$ and refer to Ω as the precision matrix, or the concentration matrix, of the multivariate normal distribution in (3.5).

This brief introduction to the multivariate normal distribution is relevant to conditional dependency networks because of the following astonishing fact: If X has a multivariate normal distribution with precision matrix Ω, then the pairs of entries of X that are conditionally dependent are precisely those pairs corresponding to nonzero entries of Ω. In other words, whenever data have a multivariate normal distribution, then the precision matrix may be viewed as an adjacency matrix for the dependence network, at least as far as the positions of the zero and nonzero entries are concerned. We close this chapter with a simple illustration of this phenomenon using three-dimensional multivariate normal observations.

3.5.2 An Illustration of Conditional Independence

Here, we consider a multivariate normal example in a 3-dimensional setting, which is the smallest nontrivial case. Observations X are generated from a multivariate normal distribution having covariance and precision matrices

$$\Sigma = \begin{pmatrix} 3 & 1 & -2 \\ 1 & 3 & -2 \\ -2 & -2 & 4 \end{pmatrix} \text{ and } \Omega = \Sigma^{-1} = \frac{1}{4}\begin{pmatrix} 2 & 0 & 1 \\ 0 & 2 & 1 \\ 1 & 1 & 2 \end{pmatrix}, \tag{3.6}$$

respectively. Recall that for any i and j, the ith and jth entries of X are conditionally independent given all the other entries if and only if $\Omega_{ij} = 0$. Thus, (3.6) leads to the dependence network of Fig. 3.5.

To understand why conditional independence is often more interesting scientifically than independence, consider that spurious correlations can often disappear when we consider conditional independence. For example, we expect a strong positive correlation between monthly ice cream sales and monthly drowning deaths in a coastal city in the northern United States. This correlation exists not because of any causal relationship between drownings and ice cream, but rather because each variable is also positively correlated with another variable, average monthly temperature: Warmer months result in both more ice cream sales and more drownings, the latter simply because more people swim in the warmer months. The key to this example is the fact that for a fixed monthly temperature–that is, if we condition on monthly temperature—there is no additional information about drownings contained in ice cream sales. Therefore, if we number ice cream sales, drownings, and temperature as variables 1, 2, and 3, respectively, we see that these variables have exactly the conditional dependence relationships depicted in Fig. 3.5.

Figure 3.6 displays values of X_1 and X_2 resulting from generating 10,000 observations from the 3-dimensional multivariate normal distribution centered at $(0, 0, 0)$ and with covariance given by (3.6). Since X_1 and X_2 are conditionally independent given X_3, we see their correlation essentially disappear when we look at subsets of points for which X_3 is roughly constant. On the other hand, the overall positive correlation between X_1 and X_2 is evident when we observe all 10,000 points.

In a statistical analysis of these 10,000 data points, our goal might be to "learn the network" of conditional independence relationships. This is exactly what the method described by (Friedman et al. 2008) does, and it is this method that resulted in the undirected network of Fig. 3.4. There are many other methods that accomplish this goal in the growing statistical literature on this topic. Most, though not all, of this

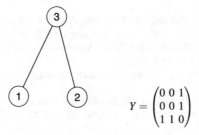

Fig. 3.5 This 3-node dependency network and its adjacency matrix correspond to the precision matrix Ω of (3.6). The off-diagonal entries of Y are nonzero exactly when the corresponding off-diagonal entries of Ω are nonzero

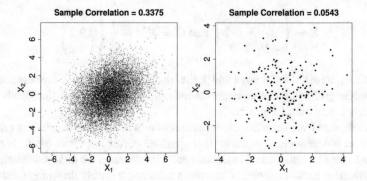

Fig. 3.6 Two-dimensional pairwise scatterplots of variables X_1 and X_2, which are generated specifically to have a positive correlation but a zero *conditional* correlation. On the left, all $10,000$ points show that the sample correlation is close to the true value of $1/3$. On the right, we see only the subset of points for which X_3 is close to zero—to be precise, they are the points for which $|X_3| < 0.05$—and the sample correlation basically disappears

work uses the convenient theory of the multivariate normal distribution as described above. Interested readers may use the references in this article as starting points to explore this literature further if they so choose.

References

Butts, C. T. (2008), 'A relational event framework for social action', *Soc. Meth.* **38**(1), 155–200.

Caimo, A. & Friel, N. (2014), 'Bergm: Bayesian exponential random graphs in R', *Journal of Statistical Software* **61**(2), 1–25.

Erdős, P. & Rényi, A. (1959), 'On random graphs I.', *Publicationes Mathematicae (Debrecen)* **6**, 290–297.

Fellows, I. E. & Handcock, M. S. (2013), 'Analysis of partially observed networks via exponential-family random network models', arXiv:1303.1219.

Friedman, J., Hastie, T. & Tibshirani, R. (2008), 'Sparse inverse covariance estimation with the graphical lasso', *Biostatistics* **9**, 432–441.

Gilbert, E. N. (1959), 'Random Graphs', *Ann. Math. Statist.* **30**(4), 1141–1144. https://doi.org/10.1214/aoms/1177706098

Goldenberg, A., Zheng, A., Fienberg, S. & Airoldi, E. (2009), 'A survey of statistical network models', *Foundations and Trends® in Machine Learning* **2**(2), 129–233.

Groendyke, C., Welch, D. & Hunter, D. R. (2012), 'A network-based analysis of the 1861 Hagelloch measles data.', *Biometrics* **68**, 755–765.

Handcock, M. S. & Gile, K. (2010), 'Modeling social networks from sampled data', *Annals of Applied Statistics* **4**, 5–25.

Handcock, M. S., Hunter, D. R., Butts, C. T., Goodreau, S. & Morris, M. (2008), 'statnet: Software tools for the representation, visualization, analysis and simulation of network data', *Journal of Statistical Software* **24**.

Handcock, M. S., Hunter, D. R., Butts, C. T., Goodreau, S. M., Krivitsky, P. N. & Morris, M. (2017), *ergm: Fit, Simulate and Diagnose Exponential-Family Models for Networks*, The Statnet Project (http://www.statnet.org). R package version 3.8.0.https://CRAN.R-project.org/package=ergm

Handcock, M. S., Raftery, A. E. & Tantrum, J. M. (2007), 'Model-based clustering for social networks (with discussion)', *Journal of the Royal Statistical Society, Series A* **170**, 301–354.

Hanneke, S., Fu, W. & Xing, E. P. (2010), 'Discrete temporal models of social networks', *Electronic Journal of Statistics* **4**, 585–605.

Hoff, P. D., Raftery, A. E. & Handcock, M. S. (2002), 'Latent space approaches to social network analysis', *Journal of the American Statistical Association* **97**(460), 1090–1098.

Hummel, R. M., Hunter, D. R. & Handcock, M. S.(2012), 'Improving simulation-based algorithms for fitting ERGMs', *Journal of Computational and Graphical Statistics* **21**(4), 920–939.

Hunter, D. R., Krivitsky, P. N. & Schweinberger, M. (2012), 'Computational statistical methods for social network models', *Journal of Computational and Graphical Statistics* **21**, 856–882.

INRA & Leger, J.-B. (2015), *blockmodels: Latent and Stochastic Block Model Estimation by a 'V-EM' Algorithm*. R package version 1.1.1. https://CRAN.R-project.org/package=blockmodels

Kolaczyk, E. D. (2009), *Statistical Analysis of Network Data: Methods and Models*, Springer.

Kolaczyk, E. D. (2017), *Topics at the Frontier of Statistics and Networks Analysis*, Cambridge University Press.

Krivitsky, P. N. & Handcock, M. S. (2014), 'A separable model for dynamic networks', *Journal of the Royal Statistical Society: Series B* **76**(1), 29–46.

Morris, M., Handcock, M. S. & Hunter, D. R. (2008), 'Specification of exponential-family random graph models: Terms and computational aspects', *Journal of Statistical Software* **24**.

Perry, P. O. & Wolfe, P. J. (2013), 'Point process modelling for directed interaction networks', *Journal of the Royal Statistical Society: Series B* **75**(5), 821–849.

R Core Team (2018), *R: A Language and Environment for Statistical Computing*, R Foundation for Statistical Computing, Vienna, Austria. https://www.R-project.org/

Ripley, R., Boitmanis, K., Snijders, T. A. B. & Schoenenberger, F. (2017), *RSiena: Siena - Simulation Investigation for Empirical Network Analysis*. R package version 1.2-3. https://CRAN.R-project.org/package=RSiena

Sachs, K., Perez, O., Pe'er, D., Lauffenburger, D. A. & Nolan, G. P. (2005), 'Causal protein-signaling networks derived from multiparameter single-cell data', *Science* **308**(5721), 523–529. http://science.sciencemag.org/content/308/5721/523

Shalizi, C. R. & Rinaldo, A. (2013), 'Consistency under sampling of exponential random graph models', *Annals of Statistics* **41**(2), 508–535.

Snijders, T. A. B. (2017), 'Stochastic actor-oriented models for network dynamics', *Annual Review of Statistics and Its Application* **4**, 343–363.

Snijders, T. A. B. & Nowicki, K. (1997), 'Estimation and prediction for stochastic blockmodels for graphs with latent block structure', *Journal of Classification* **14**, 75–100.

van Duijn, M. A. J., Snijders, T. A. B. & Zijlstra, B. J. H. (2004), 'p2: a random effects model with covariates for directed graphs', *Statistica Neerlandica* **58**, 234–254.

Wang, P., Robins, G. & Pattison, P. (2006), 'PNet: A program for the simulation and estimation of exponential random graph models', *University of Melbourne*.

Wasserman, S. & Pattison, P. (1996), 'Logit models and logistic regressions for social networks: I. An introduction to Markov graphs and p*', *Psychometrika* **61**(3), 401–425.

Žiberna, A. (2018), *blockmodeling: Generalized and Classical Blockmodeling of Valued Networks*. R package version 0.3.1.

Chapter 4
The Rank-One and the Preferential Attachment Paradigm

Steffen Dereich

Abstract Starting from the late 1990s, the analysis of complex networks has attracted significant attention within the scientific community. With the birth of social networks, information became available on personal interaction on large scales which immediately raised the question about the nature of this interaction. Mathematically one asks for models that have similar features as real-world complex networks. In particular, one wants to understand the reasons for the prevalent artefacts of complex networks. In this article, we review some classical models with a focus on two paradigms, rank-one models and preferential attachment.

4.1 Introduction

Complex networks are omnipresent. They appear for instance as

- social networks (the interaction within a large group of individuals),
- the Internet (the routers connected by the physical edges),
- the World Wide Web (the Web sites with hyperlinks)
- contact/infection networks,
- power grids, and
- predator-prey networks.

Mathematically, one represents a network as a large directed or undirected graph. A standard theoretical approach is to model a network by a sequence of random graphs $(G_n)_{n \in \mathbb{N}}$ and, in the following, G_n will denote a random (multi)graph with $[n] := \{1, \ldots, n\}$ being the set of nodes unless stated otherwise. Further E_n will denote the random set of edges of the graph G_n.

We give/recall two phenomena that can be observed in real-world networks and that have played an important role in the theory on complex networks.

S. Dereich (✉)
Institute of Mathematical Stochastics, Westfälische-Wilhelms Universität Münster, Münster, Germany
e-mail: steffen.dereich@wwu.de

© Springer Nature Switzerland AG 2019   43
F. Biagini et al. (eds.), *Network Science*, https://doi.org/10.1007/978-3-030-26814-5_4

The *small world phenomenon* refers to the fact that despite of the large size of a network most of its constituents have a surprisingly small graph (hop-count) distance. Travers and Milgram (1969) published in 1969 a study in which 296 individuals in Omaha (Nebraska) were asked to send a postcard to a stockbroker in Boston via a chain of friends, acquaintances, or relatives. 64 letters reached the target person and the authors observed a chain length with in average 4.4 to 5.7 intermediaries depending on the group that received the original postcard. The phenomenon was further popularized by John Guare's 1990 play "Six Degrees of Separation." Recent studies suggest that the distances in social networks are even smaller: The research project (Backstrom et al. 2011) observed in 2011 that two randomly picked users of Facebook are linked by an average of 3.74 intermediary friends only.

The second central phenomenon is the *scale-free* nature of networks meaning that the relative number of vertices of a prescribed degree k decays like a polynomial in k. For instance, Faloutsos et al. (1999) carried out an empirical analysis of the Internet and observed that the relative number of vertices of degree k is well described by a monomial $k^{-\tau}$ with τ being a number between 2,15 and 2,2 depending on the datasets used. In view of a complex network model $(G_n)_{n\in\mathbb{N}}$ we call a distribution $\mu = (\mu_k)_{k\in\mathbb{N}_0}$ on $\mathbb{N}_0 = \mathbb{N} \cap \{0\}$ with

$$\lim_{n\to\infty} \frac{1}{n} \#\{m \in [n] : \deg_{G_n}(m) = k\} = \mu_k, \text{ in probability}, \tag{4.1}$$

for all $k \in \mathbb{N}_0$, *asymptotic degree distribution* of the network $(G_n)_{n\in\mathbb{N}}$. If the asymptotic degree distribution μ satisfies

$$\mu_k = k^{-\tau+o(1)} \quad \text{as } k \to \infty,$$

for a $\tau > 1$, we say that (G_n) (and also μ) has power law exponent τ.

Complex networks have been the topic of several recent introductory mathematical books such as the textbooks by Bollobás (2001), Chung and Lu (2006), Durrett (2010), and van der Hofstad (2016). In this short note, we focus on two paradigms of models that have been thoroughly studied in recent years, the rank-one and the preferential attachment paradigm. The article intends to be an easy to read first introduction into these two classes of networks.

4.2   Network Models

An abundance of network models has been designed and analyzed in previous years, and we will focus on certain particular examples that are archetypical for larger classes of models.

4.2.1   The Rank-One Paradigm

The configuration model of Bollobás (2001) is the simplest model featuring heavy-tailed degree distributions. We first introduce a configuration graph.

Definition 4.1 Let $n \in \mathbb{N}$ and $\mathbf{d} = (d_k^{(n)})_{k=1,\dots,n} \in \mathbb{N}_0^n$ be such that

$$\sum_{k=1}^n d_k^{(n)} \in 2\mathbb{N}_0. \tag{4.2}$$

A *configuration graph* G_n is generated by attaching to each node $k = 1, \dots, n$, $d_k^{(n)}$ half-edges and by joining the half-edges uniformly at random, see Fig. 4.1 for an illustration. Formally, a uniform pairing of the set

$$\{(k, j) : k = 1, \dots, n, \ j = 1, \dots, d_k^{(n)}\}$$

is picked and the random multigraph $G_n = ([n], E_n)$ is formed by assigning each unordered pair $\langle (k, j), (k', j') \rangle$ an edge $\langle k, k' \rangle$ in E_n. We denote the distribution of the corresponding multigraph by $\mathrm{CM}_n(\mathbf{d})$.

Next, we introduce the corresponding network model.

Definition 4.2
1. A *configuration network model* is a sequence $(G_n)_{n \in \mathbb{N}}$ of configuration graphs with $[n]$ being the set of nodes of G_n, and we denote the respective degrees by $d_k^{(n)} = \deg_{G_n}(k)$ which is deterministic by definition.
2. If a configuration model (G_n) has asymptotic degree distribution μ with finite pth moment $(p \geq 1)$ and

$$\frac{1}{n} \sum_{k=1}^n (d_k^{(n)})^p \to \int w^p \, d\mu(w),$$

   then we call μ L^p-asymptotic degree distribution of (G_n).

A closely related model is the Norros-Reittu model (Norros and Reittu 2006).

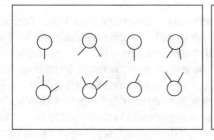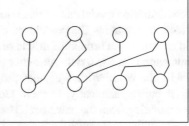

Fig. 4.1 A configuration graph. First half-edges are attached and then uniformly paired

Definition 4.3 (The Norros-Reittu model) Let ν be a distribution on $(0, \infty)$. The *Norros-Reittu model* with weight (or capacity) distribution ν is built in two steps:

1. The weights $(w_k)_{k\in\mathbb{N}}$ form a sequence of independent ν-distributed random variables and $\ell_n := \sum_{k=1}^{n} w_k$ for $n \in \mathbb{N}$.
2. Given $(w_k)_{k\in\mathbb{N}}$, each random graph G_n ($n \in \mathbb{N}$) is obtained by independently placing for each unordered pair $\langle i, j \rangle$ of distinct vertices $i, j \in [n]$ an edge $\langle i, j \rangle$ with probability

$$1 - e^{-w_i w_j / \ell_n}.$$

A sequence $(G_n)_{n\in\mathbb{N}}$ of random graphs constructed in such a way is called *Norros-Reittu graph* with weight distribution ν, briefly NR(ν)-model.

In the definition of the Norros-Reittu model, we do not say anything about the dependencies for different graph sizes since this will not play any role in our considerations. We note that the original definition in Norros and Reittu (2006) is based on a particular dynamic building rule and it allows multiple edges between the same pair of vertices.

The configuration and Norros-Reittu model are closely related since at least on an informal level in the configuration model

$$\mathbb{P}(i \overset{G_n^{CM}}{\leftrightarrow} j) \sim \frac{d_i^{(n)} d_j^{(n)}}{\sum_{k=1}^{n} d_k^{(n)}}$$

and in the Norros-Reittu model

$$\mathbb{P}(i \overset{G_n^{NR}}{\leftrightarrow} j | (w_k)_{k\in\mathbb{N}}) \sim \frac{w_i w_j}{\ell_n}.$$

Here we write $a_n \sim b_n$ if $\lim_{n\to\infty} a_n/b_n = 1$ for two given sequences $(a_n)_{n\in\mathbb{N}}$ and $(b_n)_{n\in\mathbb{N}}$ of nonnegative reals. The quantity $d_j^{(n)}$, resp. w_j, gives each node a certain weight and the likeliness to have a link between two vertices is approximately bilinear in the two weights. This is the main common feature of rank-one network models.

4.2.2 The Preferential Attachment Paradigm

The configuration model can be tuned to arbitrary degree sequences. But it does not provide insights about the prevalence of heavy-tailed degree distributions in real-world networks. An influential article of Barabási and Albert (1999) argues that this phenomenon is due to a rich-get-richer phenomenon in the network formation. As a model for the World Wide Web, they propose a sequence of random graphs $(G_n)_{n\in\mathbb{N}}$ which is built dynamically: In each step a new node is added together with a new edge emanating from the new node. The new edge connects randomly to one of the old nodes with a preference for nodes with high degree. A first rigorous definition of the model is given in Bollobás et al. (2001). A *preferential attachment model* with constant outdegree $m = 1$ is defined as a sequence of random (multi)graphs $(G_n)_{n \in \mathbb{N}}$ such that

1. G_1 is the graph with one node 1 and one loop (an edge linking 1 with itself),
2. G_{n+1} is obtained from G_n by adding the node $n + 1$ and by insertion of a new edge emanating from $n + 1$ to a random node in $\{1, \ldots, n + 1\}$ with conditional probability

$$\mathbb{P}(n + 1 \to k | G_n) = \tfrac{1}{2n+1}(\deg_{G_n}(k) + \mathbf{1}\{k = n + 1\}),$$

where $\deg_{G_n}(k)$ denotes the degree of k in the graph G_n and $\{n + 1 \to k\}$ denotes the event that the $n + 1$st node connects to $k \in \{1, \ldots, n + 1\}$.

Note that every new vertex establishes exactly one new edge and we briefly call $(G_n)_{n \in \mathbb{N}}$ PA$_1$-model. In the literature, various modifications and generalizations are analyzed. For instance, a network model $(\bar{G}_n)_{n \in \mathbb{N}}$ with fixed outdegree $m \in \{2, 3, \ldots\}$ is obtained by taking a PA$_1$-model $(G_n)_{n \in \mathbb{N}}$ and forming $\bar{G}_n$ by identifying vertices $(k - 1)m + 1, \ldots, km$ in G_{nm} for all $k = 1, \ldots, n$. We briefly call $(\bar{G}_n)$ PA$_m$-model, see also Bollobás et al. (2001) or Hofstad (2016) for further generalizations.

A model which can be adjusted to general asymptotic degree sequences is defined in Dereich and Mörters (2009).

Definition 4.4 Let $f : \mathbb{N}_0 \to (0, \infty)$ be a concave function with $f(0) \le 1$ and $f(1) - f(0) < 1$, a so called *attachment rule*. A PA(f)-*network* is a sequence of random directed graphs $(G_n)_{n \in \mathbb{N}}$ formed according to the following rules:

1. G_1 is the graph with the single node 1 and no edges.
2. Given G_n, G_{n+1} is formed from G_n by adding the node $n + 1$ and by independent insertion of a directed edge $n + 1 \to k$ for each $k = 1, \ldots, n$ with probability

$$\mathbb{P}(n + 1 \to k | G_n) = \frac{f(\text{indeg}_{G_n}(k))}{n}.$$

Here $\text{indeg}_{G_n}(k)$ denotes the indegree of node k in the graph G_n. Note that the graph is directed with all edges pointing from younger to older vertices.

The PA(f)-model has a limiting indegree distribution.

Theorem 4.1 *Let f be an attachment rule and $(G_n)_{n \in \mathbb{N}}$ a PA(f)-model. Then, (G_n) has L^1-asymptotic indegree distribution $\mu = (\mu_k)_{k \in \mathbb{N}_0}$ given by*

$$\mu_k = \frac{1}{1 + f(k)} \prod_{l=0}^{k-1} \frac{f(l)}{1 + f(l)},$$

that means that for every $k \in \mathbb{N}_0$

$$\lim_{n \to \infty} \frac{1}{n} \#\{j \in [n] : \text{indeg}_{G_n}(k)\} = \mu_k, \quad in \ probability,$$

μ *has finite first moment and*

$$\lim_{n \to \infty} \frac{1}{n} \# E_n = \lim_{n \to \infty} \frac{1}{n} \sum_{k=1}^{n} \text{indeg}_{G_n}(k) = \int w \, d\mu(w), \quad in \ probability.$$

For the classical variant of preferential attachment, such a statement goes back to Bollobás et al. (2001) and this statement can be found in Dereich and Mörters (2009) [Thm. 1.1]. By concavity of the attachment rule, the limit

$$\gamma := \lim_{n \to \infty} \frac{f(n)}{n}$$

exists and as a consequence of the theorem the PA(f)-model has an indegree power law exponent $\tau = 1 + \frac{1}{\gamma}$ that is

$$\mu_k = k^{-(1+\frac{1}{\gamma})+o(1)}, \quad \text{as } k \to \infty.$$

4.3   The Network as Seen by a Typical Node

The structure of a large network is best understood by adopting the viewpoint of a uniformly chosen node. For most complex network models, the local neighborhood seen by a uniform node converges in the sense of Benjamini-Schramm to a random graph and key properties of the complex network relate to properties of the corresponding graph limit.

Benjamini-Schramm convergence

To introduce the concept of Benjamini-Schramm convergence (Benjamini and Schramm 2001), we need some more notation. A *rooted graph* is a graph together with a designated node, the *root*. For a rooted graph, say (g, o) with o denoting the root, and $R \in \mathbb{N}$, we write $\mathcal{N}_R(g, o)$ for the R-neighborhood of (g, o), i.e., the rooted subgraph that is obtained from (g, o)

- by removing all nodes that cannot be reached from o by passing through less or equal to R edges and
- by removing all edges that attach to a previously removed node.

Further we say that two rooted graphs are equivalent if relabeling the nodes transforms one graph into the other and takes one root to the other.

Definition 4.5 (Benjamini-Schramm convergence)

1. A sequence of rooted random graphs $(G_n, \rho_n)_{n \in \mathbb{N}}$ (with arbitrary sets of nodes) is said to *locally converge* to a rooted random graph (H, ρ) iff for every $R \in \mathbb{N}$ and finite rooted graph (g, o)

$$\lim_{n \to \infty} \mathbb{P}\big(\mathcal{N}_R(G_n, \rho_n) \text{ is equiv. to } (g, o)\big) = \mathbb{P}\big(\mathcal{N}_R(H, \rho) \text{ is equiv. to } (g, o)\big).$$

2. A sequence of nonempty finite random graphs $(G_n)_{n \in \mathbb{N}}$ is said to *locally converge* to a rooted random graph (H, ρ) iff for an independent uniform node ρ_n of G_n one has local convergence (G_n, ρ_n) to (H, ρ). In that case, we call (H, ρ) *idealized neighborhood* of the graph model $(G_n)_{n \in \mathbb{N}}$.

Remark 4.1 Roughly speaking, Benjamini-Schramm convergence means that the probability that a neighborhood is equivalent to a particular pattern (formally a rooted graph) (g, o) converges for every possible pattern and for every size of the neighborhood.

To state the convergence results for the above models, we first need to define the corresponding random rooted graphs that appear as limit. For the configuration model, the limit is a Galton-Watson tree with a modified offspring distribution of the root.

Let μ be a distribution on $\mathbb{N}_0$ with finite first moment. We consider the *size biased distribution* reduced by one that is $\mu^* = (\mu_k^*)_{k \in \mathbb{N}_0}$

$$\mu_k^* = \frac{(k + 1)\mu_{k+1}}{\sum_{l=1}^{\infty} l \mu_l}.$$

In the case where $\mu = \delta_0$, we set $\mu^* = \delta_0$. We denote by $\text{GWP}^*(\mu)$ a Galton-Watson process in Ulam-Harris notation for which the root $\rho = \epsilon$ generates offspring with distribution μ and all other vertices generate independently μ^*-distributed offspring.

To be precise, let $\mathbb{N}^* := \bigcup_{\ell \in \mathbb{N}_0} \mathbb{N}^\ell$ be the set of all finite words over $\mathbb{N}$ including the empty word ϵ and let $(Z_w : w \in \mathbb{N}^*)$ be independent random variables with

$$Z_\epsilon \sim \mu \text{ and } Z_w \sim \mu^*, \text{ for } w \in \mathbb{N}^* \backslash \{\epsilon\}.$$

The random set of vertices V of a $\text{GWP}^*(\mu)$ is obtained by iteratively adding words of length $0, 1, \ldots$ according to the following rule: the root ϵ is in V and a concatenation wj of the word $w \in \mathbb{N}^*$ and the letter $j \in \mathbb{N}$ is in V if

$$w \in V \text{ and } j \leq Z_w.$$

Furthermore, each $wj \in V$ is linked by an undirected edge to w.

Theorem 4.2 *A CM(**d**)-network $(G_n)_{n \in \mathbb{N}}$ with L^1-asymptotic degree distribution μ has idealized neighborhood $(\text{GWP}^*(\mu), \epsilon)$.*

The idealized neighborhood of the PA(f)-model

Next, we introduce the idealized neighborhood of the PA(f)-model for an arbitrary attachment rule f. It is a multitype truncated branching random walk.

Let $(S_n)_{n \in \mathbb{N}_0}$ be independent random variables with each S_n being Exp($f(n)$)-distributed and let Π^+ denote the simple, locally finite point process given by

$$\Pi^+ = \sum_{k=0}^{\infty} \delta_{S_0 + \ldots + S_k}.$$

It is the point process on $(0, \infty)$ with inter-arrival times $(S_n)_{n \in \mathbb{N}_0}$. The point process Π^+ has intensity measure

$$\mathbb{E}[f(Z_t)]\,dt,$$

where $Z_t = \Pi^+((0, t]) = \sum_{k=0}^{\infty} \mathbf{1}\{S_0 + \ldots + S_k \leq t\}$ ($t \geq 0$) is the counting process associated with the point process Π^+. It is a pure birth Markov jump process with birth rate equal to f. Further we denote by Π^- the Poisson point process on $(-\infty, 0)$ with intensity measure

$$e^t\,\mathbb{E}[Z_{-t}]\,dt.$$

We now introduce the truncated multitype branching random walk TBRW(f) on $(-\infty, 0]$ with typespace $\{\ell\} \cup (0, \infty)$. Its root is a node ρ with $-$Exp(1)-distributed location $X(\rho)$ and type $T(\rho) = \ell$. The next generation is formed by taking independent copies of Π^- and Π^+ (also independent of ρ), say $\Pi^{\rho,-}$ and $\Pi^{\rho,+}$, and by placing for each point $t \in \Pi^{\rho,-} \cup \Pi^{\rho,+}$ with $X(\rho) + t \leq 0$ a node v (descendant) with

$$X(v) = X(\rho) + t \text{ and } T(v) = \begin{cases} \ell, & \text{if } t > 0, \\ -t, & \text{if } t < 0. \end{cases}$$

All descendants are connected by edges to their mother ρ. Generally, ℓ-type vertices give independently offspring according to the above rules. The corresponding rule is illustrated in the first illustration of Fig. 4.2.

It remains to explain how vertices v with type $\tau := T(v) \in (0, \infty)$ generate offspring. For such a node one takes an independent copy of Π^-, say $\Pi^{-,v}$ together with an independent point process $\Pi^{+,v}$ being distributed according to

$$\mathscr{L}(\Pi^+ \setminus \{\tau\} \mid \Pi^+(\{\tau\}) = 1).$$

Note that the latter term refers to the conditional probability of $\Pi^+ \setminus \{\tau\}$ given that $\Pi^+(\{\tau\}) = 1$. The set of descendants is then formed as in the case of ℓ-type vertices and all its constituents are connected to their mother. The corresponding random graph including the locations and types will be denoted by TBRW(f).

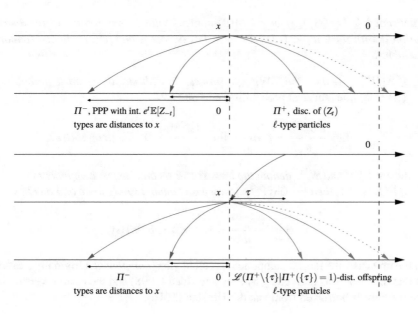

Fig. 4.2 Reproduction mechanism for the vertices of a TBRW(f).

In the case that the attachment rule is *affine*, the distribution

$$\mathscr{L}(\Pi^+\setminus\{\tau\}|\Pi^+(\{\tau\}) = 1)$$

does not depend on the choice of $\tau \in (0, \infty)$ and it equals the distribution of

$$\bar{\Pi}^+ := \sum_{k=1}^{\infty} \delta_{S_1+\ldots+S_k}.$$

Hence in that case, we deal with a *two*-type truncated branching random walk which simplifies matters.

Theorem 4.3 *A* PA(f)-*model* $(G_n)_{n\in\mathbb{N}}$ *has idealized neighborhood* (TBRW(f), ρ).

Although not being stated like that, Thm. 4.3 is a consequence of the analysis of certain exploration schemes carried out in Dereich and Mörters (2013).

4.4   The Giant Component

At least on an informal level, the Benjamini-Schramm limit allows to deduce the asymptotic size of the largest component.

Theorem 4.4 *Let* $(G_n)_{n \in \mathbb{N}}$ *be a* CM(**d**)-*network with* L^1-*asymptotic degree distribution* μ *with* $\mu_0 + \mu_2 < 1$. *Let* D, *resp.* D^*, *denote* μ *and* μ^*-*distributed random variables.*

1. *If* $\mathbb{E}[D^*] > 1$, *then the* $\mathrm{GWP}^*(\mu)$-*process is with strictly positive probability* $\zeta > 0$ *infinite* (*survival probability*) *and one has*

$$\lim_{n \to \infty} \frac{\#\mathscr{C}_n^{\max}}{n} = \zeta \quad and \quad \lim_{n \to \infty} \frac{\#\mathscr{C}_n^{2\mathrm{nd}}}{n} = 0, \quad in \ probability,$$

where $\mathscr{C}_n^{\max}$ *and* $\mathscr{C}_n^{2\mathrm{nd}}$ *denote the largest and second largest component of* G_n.
2. *If* $\mathbb{E}[D^*] \leq 1$, *then the* $\mathrm{GWP}^*(\mu)$-*process is almost surely finite and one has*

$$\lim_{n \to \infty} \frac{\#\mathscr{C}_n^{\max}}{n} = 0, \quad in \ probability.$$

In the first case, the largest component is called *giant component*. Originally a result in that direction was derived by Molloy and Reed (1998) and the current version of the theorem is borrowed from van der Hofstad (2016).

Remark 4.2 The survival probability can be represented with the help of the generating function of the distribution μ^* that is

$$g(x) = \mathbb{E}[x^{D^*}] = \sum_{k=0}^{\infty} \mu_k^* x^k, \quad x \in [0, 1].$$

The extinction probability of a Galton-Watson process with offspring distribution μ^* is the smallest fixed point ξ of g and the asymptotic size ζ of the giant component satisfies

$$\zeta = \sum_{k=1}^{\infty} \mu_k (1 - \xi^k).$$

We note that Thm. 4.4 holds similarly for the Norros-Reittu model. Let $(G_n)_{n \in \mathbb{N}}$ be a NR(ν)-network for an integrable distribution ν. Here the Benjamini-Schramm limit is a modified Galton-Watson process with the root generating offspring according to

$$\mu_k = \mathbb{E}\Big[\frac{w^k}{k!} e^{-w}\Big]$$

and all other vertices generating offspring according to

$$\mu_k^* = \frac{1}{\mathbb{E}[w]} \mathbb{E}\Big[\frac{w^{k+1}}{k!} e^{-w}\Big],$$

where w is a ν-distributed random variable. In particular, a giant component exists if and only if $\mathbb{E}[w^2] > \mathbb{E}[w]$ since the first moment of μ^* satisfies

$$\sum_{k=1}^{\infty} k\mu_k^* = \frac{1}{\mathbb{E}[w]}\sum_{k=1}^{\infty}\mathbb{E}\Big[w^2\frac{w^{k-1}}{(k-1)!}e^{-w}\Big] = \frac{1}{\mathbb{E}[w]}\mathbb{E}\Big[w^2\sum_{k=1}^{\infty}\frac{w^{k-1}}{(k-1)!}e^{-w}\Big] = \frac{\mathbb{E}[w^2]}{\mathbb{E}[w]}.$$

The Norros-Reittu graph is a particular randomized inhomogeneous random graph whose detailed analysis can be found in Bollobás et al. (2007).

Theorem 4.5 is analogously true for the preferential attachment model (Dereich and Mörters 2013).

Theorem 4.5 *Let $(G_n)_{n\in\mathbb{N}}$ be a PA(f)-network for an attachment rule f and let ζ denote the probability that TBRW(f) is infinite. Then,*

$$\lim_{n\to\infty}\frac{\#\mathscr{C}_n^{\max}}{n} = \zeta \text{ and } \lim_{n\to\infty}\frac{\#\mathscr{C}_n^{2nd}}{n} = 0, \text{ in probability.}$$

Remark 4.3 Analytically tractable expressions for the survival probability ζ are not known for PA(f)- networks. However, one can characterize the cases in which a giant component exists. Let Γ^ℓ denote the intensity measures of the point processes $\Pi^+ \cup \Pi^-$ on $\mathbb{R}$, i.e.,

$$\Gamma^\ell(dt) = (\mathbf{1}_{\{t>0\}}\,\mathbb{E}[f(Z_t)] + \mathbf{1}_{\{t<0\}}\,e^t\,\mathbb{E}[f(Z_{-t})])\,dt.$$

Further let for $\tau \in [0,\infty]$, Γ^τ denote the intensity measure of the point process with distribution $\mathscr{L}(\Pi^- \cup \Pi^+\backslash\{\tau\}|\Pi^+(\{\tau\}) = 1)$. (Note that the conditioning on $\Pi^+(\{\tau\}) = 1$ for $\tau = \infty$ can be defined by taking limits.) We let $\mathscr{S} = \{\ell\} \cup [0,\infty]$ and consider for $\alpha \in (0,1)$ the operator

$$A_\alpha g(\tau) = \int_{\mathbb{R}} g(t)e^{-\alpha t}\,d\Gamma^\tau(t)$$

on the Banach space $\mathbf{C}(\mathscr{S})$ of bounded and continuous functions. Then, the network *does not* have a giant component if and only if there exists $\alpha \in (0,1)$ such that A_α is a well-defined operator from $\mathbf{C}(\mathscr{S})$ into itself with spectral radius $\rho(A_\alpha) < 1$.

In the case where f is affine, that is $f(n) = \gamma n + \beta$ with $\gamma \in [0,1)$ and $\beta \in (0,1]$ we only need to distinguish two types, say ℓ and r, and the intensities are explicit namely

$$\Gamma^\ell(dt) = \beta(\mathbf{1}_{\{t>0\}}\,e^{\gamma t} + \mathbf{1}_{\{t<0\}}\,e^{(1-\gamma)t})\,dt$$

and

$$\Gamma^r(dt) = (\mathbf{1}_{\{t>0\}}\,(\beta+\gamma)\,e^{\gamma t} + \beta\mathbf{1}_{\{t<0\}}\,e^{(1-\gamma)t})\,dt.$$

Thus, A_α is a 2×2-matrix and one gets that the $\mathrm{PA}(f)$-network has a giant component if and only if $\beta > 0$ and

$$\gamma \geq \frac{1}{2} \text{ or } \beta > \frac{(\frac{1}{2} - \gamma)^2}{1 - \gamma}.$$

For the details, we refer the reader to Dereich and Mörters (2013).

In the above theorems, we used the idealized neighborhoods to give representations for the size of the giant component. We want to stress that although the Benjamini-Schramm limits give good intuition about the characteristics of the network the proof of the results typically require significantly stronger statements than Benjamini-Schramm convergence. A common approach is to show that a stepwise *exploration* of G_n starting from a uniformly chosen node is probabilistically not distinguishable with the exploration of the idealized neighborhood for a certain number of steps depending on the size of the network n. Also it is sometimes preferable to use the particular structure of the model and to bypass the consideration of Benjamini-Schramm limits.

4.5 Distances

In this section, we consider typical distances[1] between two independent, uniformly chosen nodes from the giant component. We focus on the ultrasmall regime where typically distances scale like $\log \log n$ in the network size n and on the boundary case, the critical case. Roughly speaking, we consider the heavy-tailed case where the distribution μ has infinite second moment. In the following, we denote by $\bar{\mu}_k = \sum_{m=k}^{\infty} \mu_m$ $(k \in \mathbb{N}_0)$ the tail probabilities of μ.

For typical distances, an intriguing difference can be observed between the rank-one and the preferential attachment paradigm in the ultrasmall regime: For the preferential attachment paradigm, distances are typically twice as long as in a corresponding rank-one model. We first consider the Norros-Reittu model.

Theorem 4.6 *Let $(G_n)_{n \in \mathbb{N}}$ be a $\mathrm{NR}(\nu)$-model for an integrable distribution ν on $(0, \infty)$. Suppose that there exists $\tau \in (2, 3)$ and $c > 0$ such that for ν-distributed random variable w*

$$\mathbb{P}(w > u) = u^{1-\tau} (c + o(1)) \quad as \ u \to \infty.$$

Then, $(G_n)_{n \in \mathbb{N}}$ has an asymptotic degree distribution μ with

$$\bar{\mu}_k = (c + o(1)) k^{1-\tau} \quad as \ k \to \infty,$$

---

[1]In a graph, the distance between two nodes is the minimal number of edges a path has to cross to go from one to the other node.

and for two independent, uniformly chosen vertices V and W in the giant component of G_n, we have

$$d_{G_n}(V, W) = (2 + o(1)) \frac{\log \log n}{\log 1/(\tau - 2)}, \quad \text{with high probability.}$$

A proof of the result can be found in Dereich et al. (2012) [Prop. 6] where the upper bound of Norros and Reittu (2006) [Thm. 4.2] is complemented by an appropriate lower bound. Thus, the typical distance grows as $\log \log n$ with an additional multiplicative constant only depending on the power law exponent τ. Similar statements hold for other rank-one models. In particular, in the configuration model a much more elaborate study (Hofstad et al. 2007) derived a distributional limit theorem for the appropriately scaled distance.

The corresponding statement for the distances in the preferential attachment network is as follows.

Theorem 4.7 *Let $(G_n)_{n\in\mathbb{N}}$ be a PA(f)-model with the attachment rule f satisfying $\gamma := \lim_{n\to\infty} f(n)/n > 1/2$. Then, (G_n) has asymptotic degree distribution μ with*

$$\mu_k = k^{-\tau+o(1)},$$

where $\tau = 1 + 1/\gamma$, and for two independent, uniformly chosen vertices V and W in the giant component of G_n, we have

$$d_{G_n}(V, W) = (4 + o(1)) \frac{\log \log n}{\log 1/(\tau - 2)} \quad \text{with high probability.}$$

The proof of the theorem can be found in Dereich et al. (2012) [Prop. 4]. We should mention that the proof of the upper bound uses arguments from Dommers et al. (2010) where a corresponding upper bound is derived for a different variant of preferential attachment.

To understand the additional factor two in the preferential attachment paradigm, we give a rough sketch of the proofs of the upper bounds. A visualization can be found in Fig. 4.3.

In the proofs of the upper bounds, one shows existence of certain connecting paths. Roughly speaking, one assigns every node to either one of $(c + o(1)) \log \log n$ shells or the bulk, where $c = 1/\log 1/(\tau - 2)$. The shells are chosen in such a way that the innermost shell contains the vertices with highest weights in the Norros-Reittu network and the oldest vertices in the preferential attachment network. The weight/age decreases when moving from the innermost to the outermost shell. For the Norros-Reittu network, one shows the following:

- A uniformly chosen node is with high probability element of the bulk. If it is in the giant component, it has, with high probability, distance $\mathcal{O}(1)$ to the outermost shell.

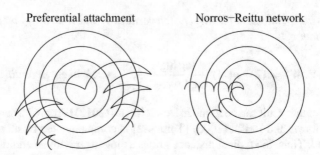

Fig. 4.3 Illustration of the connecting paths used in the proofs of the upper bounds for Norros-Reittu and preferential attachment.

- The first node found in the outermost shell features with high probability a link to the next shell and so on, so that one finds with high probability a path of the number of shells minus one leading from the outermost to the innermost shell.
- In the case that the above construction succeeds for two uniform vertices, the distance between the corresponding two vertices in the innermost shell has distance of order $\mathscr{O}(1)$, with high probability.

Hence, with high probability two uniformly chosen vertices have distance less than or equal to $(2c + o(1)) \log \log n$.

For the preferential attachment model, the proof can be achieved following the same steps with a slight modification. In the second step, one finds, with high probability, a bridge to the next shell which has one intermediary bulk node so that one finally finds connecting paths between two independent uniformly chosen vertices with length $(4c + o(1)) \log \log n$.

The classical preferential attachment model PA_m has power law exponent $\tau = 3$ which is the critical exponent. Bollobás and Riordan (2004) showed that two randomly chosen nodes have graph distance $(1 + o(1)) \log n / \log \log n$, with high probability. The same result holds for a variety of rank-one models such as the Norros-Reittu and the configuration model, with asymptotic degree distribution

$$\bar{\mu}_k = (c + o(1))k^{-2}.$$

Hence, in that case the asymptotics of the two paradigms agree including the constant.

These results raise the question whether there is an intermediate regime between the ultrasmall and critical regime in which the ratio between the typical distances interpolates between one and two. Such a behaviour can be observed for degree distributions μ with tail behavior

$$\bar{\mu}_k = k^{-2}(\log k)^{2\alpha + o(1)}, \quad \text{as } k \to \infty, \tag{4.3}$$

where $\alpha > 0$. The following results are taken from Dereich et al. (2017).

Theorem 4.8 *Let $\alpha > 0$ and $(G_n)_{n \in \mathbb{N}}$ denote the Norros-Reittu network with independent identically distributed weights $\mathbf{w} = (w_k^{(n)} : n \in \mathbb{N}, k = 1, \ldots, n)$ satisfying*

$$\mathbb{P}(w_k^{(n)} \geq u) = u^{-2}(\log u)^{2\alpha + o(1)}, \tag{4.4}$$

as $u \to \infty$. Then, (4.3) is satisfied and for two independent, uniformly chosen nodes U, V from the giant component $\mathscr{C}_n^{\max}$ of G_n, we have

$$d_{G_n}(U, V) = \left(\frac{1}{1 + 2\alpha} + o(1) \right) \frac{\log n}{\log \log n}, \quad \text{with high probability.}$$

The corresponding statement for the preferential attachment model is as follows.

Theorem 4.9 *Let $(G_n)_{n \in \mathbb{N}}$ be the preferential attachment model obtained from an attachment rule f satisfying*

$$f(k) = \frac{1}{2}k + \frac{\alpha}{2} \frac{k}{\log k} + o\left(\frac{k}{\log k} \right), \tag{4.5}$$

for some $\alpha > 0$. Then, (4.3) is satisfied and for two independent, uniformly chosen nodes U, V from the giant component $\mathscr{C}_n^{\max} \subset G_n$, we have

$$d_{G_n}(U, V) = \left(\frac{1}{1 + \alpha} + o(1) \right) \frac{\log n}{\log \log n}, \quad \text{with high probability.}$$

We note that in the parameter $\alpha > 0$ the typical distance in the preferential attachment model scales as $\frac{1 + 2\alpha}{1 + \alpha}$ times the distance in the corresponding Norros-Reittu network and in the limit $\alpha \to \infty$ one obtains the additional factor 2. Unfortunately, the proofs of these results do not provide intuition about the structural differences that cause the additional factor.

References

Backstrom, L., Boldi, P., Rosa, M., Ugander, J. & Vigna, S. (2011), 'Four degrees of separation', arXiv:1111.4570.

Barabási, A.-L. & Albert, R. (1999), 'Emergence of scaling in random networks', *Science* **286**.

Benjamini, I. & Schramm, O. (2001), 'Recurrence of distributional limits of finite planar graphs', Electron. J. Probab. 6(23), 1–13.

Bollobás, B. (2001), Random Graphs, Cambridge Studies in Advanced Mathematics, 2nd edn, Cambridge University Press, Cambridge.

Bollobás, B., Janson, S. & Riordan, O. (2007), 'The phase transition in inhomogeneous random graphs', Random Structures and Algorithms 31, 3–122.

Bollobás, B., Riordan, O., Spencer, J. & Tusnády, G. (2001), 'The degree sequence of a scale-free random graph process', Random Structures Algorithms 18(3), 279–290.

Bollobás, B. & Riordan, O. (2004), 'The diameter of a scale-free random graph', Combinatorica 24, 5–34.

Chung, F. & Lu, L. (2006), *Complex graphs and networks*, Vol. 107 of *CBMS Regional Conference Series in Mathematics*, Published for the Conference Board of the Mathematical Sciences, Washington, DC.

Dereich, S., Mönch, C. & Mörters, P. (2012), 'Typical distances in ultrasmall random networks', Adv. in Appl. Probab. 44(2), 583–601.

Dereich, S., Mönch, C. & Mörters, P. (2017), 'Distances in scale free networks at criticality', Electron. J. Probab. 22(77), 1–38.

Dereich, S. & Mörters, P. (2009), 'Random networks with sublinear preferential attachment: degree evolutions', Electron. J. Probab. 14(43), 1222–1267.

Dereich, S. & Mörters, P. (2013), 'Random networks with sublinear preferential attachment: the giant component', Ann. Probab. 41(1), 329–384.

Dommers, S., Hofstad, R. v. d. & Hooghiemstra, G. (2010), 'Diameters in preferential attachment models', Journal of Statistical Physics 139, 72–107.

Durrett, R. (2010), *Random Graph Dynamics*, Vol. 20 of *Cambridge Series in Statistical and Probabilistic Mathematics*, Cambridge University Press, Cambridge.

Faloutsos, M., Faloutsos, P. & Faloutsos, C. (1999), 'On power-law relationships of the internet topology', *SIGCOMM Comput. Commun. Rev.* **29**(4), 251–262. http://doi.acm.org/10.1145/316194.316229

Hofstad, R. v. d., Hooghiemstra, G. & Znamenski, D. (2007), 'Distances in random graphs with finite mean and infinite variance degrees', Electronic Journal of Probability 12, 703–766.

Hofstad, R. v. d. (2016), *Random Graphs and Complex Networks*, Vol. 1 of *Cambridge Series in Statistical and Probabilistic Mathematics*, Cambridge University Press.

Molloy, M. & Reed, B. (1998), 'The size of the giant component of a random graph with a given degree sequence', Combin. Probab. Comput. 7(3), 295–305.

Norros, I. & Reittu, H. (2006), 'On a conditionally Poissonian graph process', Advances in Applied Probability 38, 59–75.

Travers, J. & Milgram, S. (1969), 'An experimental study of the smal world phenomenon', Sociometry 32(4), 425–443.

Chapter 5
Systemic Risk in Networks

Nils Detering, Thilo Meyer-Brandis, Konstantinos Panagiotou
and Daniel Ritter

Abstract Systemic risk, i.e., the risk that a local shock propagates throughout a given system due to contagion effects, is of great importance in many fields of our lives. In this summary article, we show how asymptotic methods for random graphs can be used to understand and quantify systemic risk in networks. We define a notion of resilient networks and present criteria that allow us to classify networks as resilient or non-resilient. We further examine the question how networks can be strengthened to ensure resilience. In particular, for financial systems we address the question of sufficient capital requirements. We present the results in random graph models of increasing complexity and relate them to classical results about the phase transition in the Erdös-Rényi model. We illustrate the results by a small simulation study.

5.1 Introduction

One possible attempt to define *Systemic Risk* is that in case of an adverse local shock (infection) to a system of interconnected entities, a substantial part of the system, or even the whole system, finally becomes infected due to contagion effects. In an evermore connected world, systemic risk is an increasing threat in many fields of our life; examples include the epidemic spread of diseases, the collapse of financial

N. Detering (✉)
Department of Statistics and Applied Probability, University of California, California, USA
e-mail: detering@pstat.ucsb.edu

T. Meyer-Brandis · K. Panagiotou · D. Ritter
Department of Mathematics, University of Munich, Munich, Germany
e-mail: meyer-brandis@math.lmu.de

K. Panagiotou
e-mail: kpanagio@math.lmu.de

D. Ritter
e-mail: ritter@mathematik.uni-muenchen.de

© Springer Nature Switzerland AG 2019   59
F. Biagini et al. (eds.), *Network Science*, https://doi.org/10.1007/978-3-030-26814-5_5

networks, rumor spreading in social networks, computer viruses infecting servers, or breakdowns of power grids. However, as, for example, the recent financial crisis has demonstrated, traditional risk management strategies and techniques often only inadequately account for systemic risk as they predominantly focus on the single system entities and only insufficiently consider the whole system with its potentially devastating contagion effects. It is thus of great interest to develop new quantitative tools that can support the process of identifying, measuring, and managing systemic risk. This problem has been addressed in a number of papers now and the literature is still growing. One active line of research is the extension of the axiomatic approach to monetary risk measures from financial mathematics, initiated in Artzner et al. (1999), to systemic risk measures (Chen et al. 2013; Hoffmann et al. 2016, 2018; Kromer et al. 2016; Biagini et al. 2019+; Armenti et al. 2018; Feinstein et al. 2017). Another interesting analysis of systemic risk is based on an explicit modeling of the underlying network of interacting entities and potential contagion effects (Amini et al. 2015; Chong and Klüppelberg 2018; Eisenberg and Noe 2001; Gandy and Veraart 2017; Kusnetsov and Veraart 2016; Weber and Weske 2017). For a further overview of different methods and concepts to address systemic risk, the reader is referred to the two monographs Fouque and Langsam (2013) and Hurd (2016). In this chapter, we give an account of how such tools can be developed for large networks in the framework of random graph models, which allows for an explicit modeling both of the underlying network structure and of the contagion process propagating systemic risk. This area of research has been initiated by the work Gai and Kapadia (2010) and Amini et al. (2016) and further developed in Detering et al. (2015, 2016).

For a system with n entities (nodes) with labels in $[n] := \{1, \ldots, n\}$, rather than dealing with a specific deterministic network we consider a random graph model G on a given probability space $(\Omega, \mathcal{F}, \mathbb{P})$, where for each scenario $\omega \in \Omega$ the realized network is represented by its adjacency matrix $G(\omega) \in \{0, 1\}^{n \times n}$. We want to exclude self-loops from the graph and hence assume $G_{i,i} = 0$ almost surely for all $i \in [n]$. Further, if we consider undirected networks, then $G_{i,j}(\omega) = G_{j,i}(\omega)$ for all $i, j \in [n]$. If $G_{i,j}(\omega) = 1$, then we say that (i, j) is an edge of G.

The motivation for choosing such a random graph framework is twofold. From a modeling point of view, risk management deals with the uncertainty of adverse effects at a future point in time. In many situations, however, not only the future adverse shock is uncertain but also the specific network structure. While the statistical characteristics (degree distribution, . . .) of the considered network often remain stable over time, the specific configuration of edges may change. This uncertainty is then represented by a suitable random graph model. Secondly, from a mathematical point of view, the framework of random graphs allows for the application of the law of large number effects when the network size gets large. This enables the analytic derivation of asymptotic results that hold for all "typical" future realizations $G(\omega)$ of large networks (more precisely, the results hold *with high probability* [w. h. p.], that is with probability growing to 1 when n tends toward infinity). In this sense, our systemic risk management results are robust with respect to the uncertainty about the future network configuration and applicable to all "typical" networks that share the same statistical characteristics.

The random graph models we consider are characterized by the fact that each possible edge e is included *independently* with some probability p_e. Our models differ in the way the marginal probabilities are actually specified: If all p_e are equal, then we call this the *homogeneous* setting. This is the classical Erdös-Rényi model. We refer the reader to Bollobás (2001), Alon and Spencer (2016), and Janson et al. (2011) for an excellent introduction to the model and its asymptotic properties. All other cases are called *heterogeneous*, and the resulting model is often called the inhomogeneous random graph in the literature. In the latter case, the actual degree of heterogeneity has an important effect on several structural characteristics of the resulting graphs (as, e.g., the distribution of the edges in the graph, or the emergence of a core-periphery structure), and we will exploit this effect to capture realistic situations. The recent monograph Hofstad (2016) gives an extensive introduction to this heterogeneous model and its alternatives including the configuration model.

In addition to the specification of a random graph model, we explicitly model the contagion effects by which an initial local shock propagates throughout the system. The contagion processes we consider in this chapter are generalizations of the so-called *bootstrap percolation process*. The essential feature specifying the contagion process is the assumption that each node i is equipped with a threshold value $\tau_i \in \mathbb{N}$ that represents the "strength" of node i to withstand contagion effects. Given a subset $I \subset [n]$ of initially "infected" nodes , the contagion process can then be described in rounds where node $i \in [n]$ gets infected as soon as τ_i of its neighbors are infected. This contagion process then clearly ends after at most $n - 1$ rounds leading to the set of eventually infected nodes triggered by the initially infected nodes I. The essential risk indicator underlying our analysis of systemic risk is then the *final infection fraction*

$$\alpha_n := \frac{\text{number of finally infected nodes}}{n} \tag{5.1}$$

given by the number of finally infected (or defaulted) nodes triggered by the set I of initially infected nodes divided by the total number of nodes in the network.

When such a contagion process is studied on a random graph G, the final default fraction $\alpha_n(\omega)$ is a random variable that depends on the realized network $G(\omega)$, and the first main question is whether one can quantify the final default fraction. Asymptotically for large networks this question can be answered positively for the random graph models we consider, and we show that the final default fraction is given by some deterministic, analytic formula depending on the statistical network characteristics and the contagion process in the limit in probability as $n \to \infty$. So roughly speaking, when the network size n is large enough, the final default fraction can be computed analytically and it will be the same for almost all network realizations $G(\omega)$ of the random graph G (and in this sense is robust with respect to the uncertainty about the future network structure).

Based on the analysis of the final default fraction, we then develop a quantitative concept to asses the systemic riskiness of a network. More precisely, we present a mathematical criterion formulated in terms of the network statistics that characterizes whether a network is resilient or non-resilient with respect to initial shocks. Roughly speaking, a network is resilient, and thus acceptable from a systemic risk point of view, if small shocks remain small, and it is non-resilient, and thus non-acceptable from a systemic risk point of view, if any initial shock propagates to a substantial part of the system, no matter how small the initial shock is. We will see that in terms of this resilience criterion the systemic riskiness of a network heavily depends on the topology of the graph. In particular, as long as the degree sequence possesses a second moment only local effects determine whether a network is resilient or not and the absence of the so-called contagious edges in the network guarantees resilience. Here, an edge (i, j) is called *contagious* if the mere infection of node j leads to the infection of node i (or vice versa). If, on the other hand, the degree sequence has infinite second moment, a property that many real-world networks share, many global effects contribute to the contagion process, and the absence of contagious links no longer implies resilience.

Once a measure of systemic risk is introduced, the second important question is how to manage systemic risk, i.e., how to design or control a system such that it is acceptable from a systemic risk point of view. In our framework, we analyze this question in the following sense: For a given graph structure, how does one have to specify the threshold values τ_i, $i \in [n]$, such that the network becomes resilient? For example, in the context of financial networks, requirements on τ_i can be inter-preted as capital requirements imposed on a financial institution $i \in [n]$. Using above-mentioned resilience criterion, it follows immediately that for networks with finite second moment of the degree sequence the requirement $\tau_i \geq 2$, $i \in [n]$, is sufficient for resilience since this excludes contagious edges. For networks without finite sec-ond moment of the degree sequence this management rule is insufficient for securing a system and we will see that highly connected nodes need to be equipped with higher threshold values. In particular, we will characterize resilience/non-resilience in terms of a specific functional form for the threshold values, where the threshold value τ_i for node i can still basically be determined locally by only knowing the profile of node i. This striking feature is possible due to averaging effects in large random graphs and it is in contrast to other management (or allocation) rules obtained in deterministic networks that for each node can only be specified in terms of the complete network structure.

In the course of this chapter, we expose the program sketched above in gradually increasing complexity of both the underlying random graph model and the conta-gion process. In Sect. 5.2.1, we consider the homogeneous setting of the well-studied Erdös-Rényi random graph and the classical bootstrap percolation process with con-stant threshold values. In Sect. 5.2.2, to account for more realistic features of many empirically observed networks, we extend the homogeneous setting to both hetero-geneous random graphs and threshold values, which in particular allows for graphs with infinite second moment degree sequences. Finally, in Sect. 5.2.3, we focus on the modeling of financial networks where the contagion process is driven by capital endowments and exposures of the financial institutions. This contagion process rep-resents a further extension/generalization of the threshold-driven contagion process.

The results of both Sects. 5.2.2 and 5.2.3 were originally derived in Detering et al. (2015, 2016). It is the aim of this chapter to summarize this work and make our results comprehensible to a broad audience of different backgrounds.

5.2   Models of Networks and Contagion Processes

In this section, we will describe various models of random networks and contagion processes, accompanied by several results characterizing their qualitative behavior. The presentation will be such that the complexity of both the considered networks models, as well as the contagion process, increases gradually from a rather homogeneous setting to one that may resemble some realistic situations quite well.

Random Graph Models
Our random graph models have the following common characteristics. We assume that a number n of nodes with labels in $[n] := \{1, \ldots, n\}$ is given. The set of possible edges E_n consists then either of all unordered pairs $\{i, j\}$, where $i \neq j$ ("undirected graph") or all ordered pairs (i, j), where again $i \neq j$ ("directed graph"). The graph G is specified by including each possible edge e *independently* with some probability p_e. Our models differ in the way the marginal probabilities $(p_e)_{e \in E_n}$ are actually specified: If all p_e are equal, then we call this the homogeneous setting, and all other cases are called heterogeneous. In the latter case the actual degree of heterogeneity has an important effect on several structural characteristics of the resulting graphs (as, e.g., the distribution of the edges in the graph, or the emergence of a core-periphery structure), and we will exploit this effect to capture realistic situations.

Contagion Processes
The contagion processes that we consider here resemble and extend the well-studied *bootstrap percolation* process, which has its origin in the physics literature (Chalupa et al. 1979) . In the classical setting, a graph G is given and initially a subset I of the nodes is declared *infected*. We will make the assumption that each node is initially infected with some probability $\varepsilon > 0$, independently of all other nodes. The process then consists of rounds, in which further nodes may get infected. Similar as with the random graph also the infection rules that we study will become gradually more complex to represent more realistic settings. We start with the simple rule 1BP where a node becomes infected as soon as one of its neighbors becomes infected. More complex rules we study then allow for variation in the nodes' individual infection thresholds (e.g., rBP stands for the rule in which each node has threshold r) and the impact of different edges. More concrete rules will be introduced later. In all cases, we will be interested in the size of the set of eventually infected nodes. For each finite graph size n, this is a random number which depends on the realized graph configuration. However, due to averaging effects we will be able to compute the (random) fraction of eventually infected nodes $\alpha_n(\varepsilon)$ as in (5.1) as a deterministic number $\alpha(\varepsilon)$ in the limit $n \to \infty$ under some mild regularity assumptions.

Resilience to Contagion

For many applications, the spread of the initially infected set to the whole graph is of central importance. In some cases, it may be favorable if a small fraction ε of initially infected nodes spreads to a large fraction of the whole graph; in other cases, such behavior would be rather worrisome. To capture these two different kinds of possible behavior, we give the following definitions:

Definition 5.1 A network is said to be *resilient* if $\alpha(\varepsilon) \to 0$ as $\varepsilon \to 0$.

Definition 5.2 A network is said to be *non-resilient* if there exists some lower bound $\underline{\alpha} > 0$ such that $\alpha(\varepsilon) > \underline{\alpha}$ for all $\varepsilon > 0$.

Definition 5.1 characterizes a network as being resilient (to small initial infections) if the final fraction of infected nodes vanishes as the fraction of initially infected nodes ε tends to 0. In this case, small local shocks cannot cause serious harm to the system but they only impact their immediate neighborhood in the graph. On the other hand, Def. 5.2 classifies networks as non-resilient if every howsoever small initial fraction $\varepsilon > 0$ causes a positive fraction of at least $\underline{\alpha} > 0$ of eventual infections. In particular, the amplification factor $\alpha(\varepsilon)/\varepsilon$ explodes as ε becomes small and the effects are not locally confined anymore.

5.2.1 Homogeneous Setting

In this section, we study the homogeneous setting which comprises three assumptions. First, the graph is undirected, i.e., we have n nodes and the set of possible edges is $E_n := \binom{[n]}{2} = \{\{i, j\} : 1 \le i, j \le n, i \ne j\}$. Second, for every $e \in E_n$ we assume that $p_e = p$, that is, the probability that an edge is present is the same for all edges. This is a classical and well-studied model of random graphs that was first introduced in Gilbert (1959) and Erdős and Rényi (1960), and it has been investigated in great detail since then; see Bollobás (2001) for an excellent introduction. Our third and final assumption is that in the contagion process all nodes are initially infected with the same probability ε and the infection thresholds are also equal to some number $r \in \mathbb{N}$, which means that any node becomes infected as soon as (at least) r of its neighbors are infected. This infection process is well understood and treated in detail in Janson et al. (2012) and all results in this section are either special cases of results in Janson et al. (2012) or easily arise from them.

We shall use the standard notation $G_{n,p}$ for a random graph with n nodes and edge probability p as described in this section. For different choices of p, this graph shows different characteristics and can range from a very sparse, loosely connected graph to a very dense graph. In particular, note that the number $e(G_{n,p})$ of edges in $G_{n,p}$ follows a binomial distribution with parameters $|E_n| = \binom{n}{2}$ and p. Thus, the expected number of edges in $G_{n,p}$ equals $\binom{n}{2}p$, and their actual number is typically close to this value. Here, we will focus especially on the case $p = p(n) = c/n$ for some $c > 0$, as then $\mathbb{E}(e(G_{n,p})) = \frac{c}{2}(n-1) \sim cn/2$, a quantity that is linear in the number of nodes and thus most interesting for the applications that we have in mind. See Fig. 5.1 for an illustration of such a network.

A well-known structural property of the random graph $G_{n,c/n}$ that will become quite handy later is that, as $n \to \infty$, the (random) degree $\deg(i)$ of each node $i \in [n]$ converges weakly to a Poisson distribution with parameter c. Furthermore, if one considers the (random) empirical degree distribution

$$\tilde{F}_n(k) := n^{-1} \sum_{i \in [n]} \mathbf{1}\{\deg(i) \le k\}, \quad k \in \mathbb{N}_0$$

then the following statement is true (see, e.g., Hofstad (2016) [Thm. 5.12]):

Lemma 5.1 *As $n \to \infty$, $\tilde{F}_n$ converges to a Poisson distribution with parameter c.*

After having introduced the underlying random graph model for this section, we are now interested in analyzing the contagion mechanism. Recall that regarding the contagion process we assume that each node is infected initially with some probability $\varepsilon > 0$ and independently of all other nodes. Nodes that are not initially infected shall become so as soon as $r \in \mathbb{N}$ of their neighbors are infected, i.e. $\tau_i = r$ for all nodes $i \in [n]$ that are not infected initially. In the sequel, we distinguish the cases $r = 1$ and $r \ge 2$ for the infection threshold of each node. Our main focus will be to distinguish between two fundamentally different behaviors:

r = 1: Observe that in this case a node gets infected as soon as any of its neighbors is infected. In particular, if i is a node that was infected at the beginning of the process, then eventually the whole connected component containing i will become infected; hence, the behavior of the process is intimately related to the component structure of $G_{n,p}$. Here, the famous result of Erdös and Rényi (see Alon and Spencer (2016), for example) regarding a phase transition in the component structure comes to help. Let us write $L(G_{n,p})$ for the random number of nodes in a largest connected component of $G_{n,p}$.

Fig. 5.1 A typical configuration for $G_{n,p}$ with $n = 100$, $c = 4$, and $p = c/n$. Node sizes scale with the corresponding degree

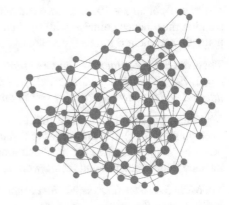

Theorem 5.1 *Let $c > 0$ and $p = c/n$. Then, as $n \to \infty$,*

- *if $c < 1$, then there exists $\kappa \in (0, \infty)$ such that $\log^{-1}(n)L(G_{n,p}) \to \kappa$ in probability.*
- *if $c > 1$, then there exists $\lambda \in (0, \infty)$ such that $n^{-1}L(G_{n,p}) \to \lambda$ in probability.*

A first important consequence of this result is that if $c > 1$, then, no matter how small $\varepsilon > 0$ is chosen, w. h. p. (i. e. with probability converging to 1 as $n \to \infty$) at least one node in the largest component will be infected and in turn at least a fraction λ of the nodes in the graph will eventually become infected. We hence derive the following result:

Theorem 5.2 *Consider the random graph model $G_{n,p}$ with $p = c/n$ and threshold $r = 1$. If $c > 1$, then the system is non-resilient.*

Regarding the case $c < 1$, it turns out that $G_{n,c/n}$ is resilient according to Def. 5.1. We do, however, need more information about the random graph than only the size of its largest component in order to conclude this. Indeed, from a heuristic point of view, the following consideration is helpful: Let $\alpha(\varepsilon) \in [\varepsilon, 1]$ be the (a priori possibly random) fraction of eventually infected nodes. Each of the eventually infected nodes is either infected from the beginning, which happens with probability ε, or otherwise it must have at least one infected neighbor. We know that the degree of each node is Poisson distributed with mean c in the limit $n \to \infty$. Since a fraction $\alpha(\varepsilon)$ of all nodes is eventually infected, for each node i that becomes infected during the process (not initially infected), the number of infected neighbors can be expected to be Poisson distributed with parameter $c\alpha(\varepsilon)$. This heuristic argument yields the identity

$$\alpha(\varepsilon) = \varepsilon + (1 - \varepsilon)\mathbb{P}(\mathrm{Poi}(c\alpha(\varepsilon)) \geq 1)$$

The fraction $\alpha(\varepsilon)$ should then be a fixed point of the function

$$f_\varepsilon(z) := \varepsilon + (1 - \varepsilon)\mathbb{P}(\mathrm{Poi}(cz) \geq 1).$$

Since f_ε is continuous, $f_\varepsilon(0) = \varepsilon > 0$, and $f_\varepsilon(1) \leq \varepsilon + (1 - \varepsilon) = 1$, there always exists at least one fixed point of f_ε within $(0, 1]$. Further, since $f_\varepsilon''(z) = -c^2(1 - \varepsilon)e^{-cz} < 0$ for all $z \in [0, \infty)$, there can only exist one solution $\hat{z}$ to $f_\varepsilon(z) = z$. This solution must hence coincide with the final fraction of infected nodes. Making this heuristic argument rigorous (compare to Thm. 5.5), we derive the following result.

Theorem 5.3 *Consider the random graph model $G_{n,p}$ with $p = c/n$ and threshold $r = 1$. Let $\hat{z}$ denote the unique fixed point of $f_\varepsilon(z)$. Then, the fraction of eventually infected nodes converges to $\alpha(\varepsilon) = \hat{z}$ in probability.*

Regarding the case $c < 1$, note that as the initial infection probability $\varepsilon \to 0$, the fixed point of $f_\varepsilon(z)$ also converges to 0 since $f_\varepsilon(z) \leq \varepsilon + (1 - \varepsilon)cz$ and hence $\hat{z} \leq \varepsilon(1 - c)^{-1}$. See Fig. 5.2a for an illustration. This means that the final fraction of infected nodes vanishes and the network is thus resilient according to Def. 5.1:

Theorem 5.4 *Consider the random graph $G_{n,p}$ with $p = c/n$ and threshold $r = 1$. If $c < 1$, then the system is resilient.*

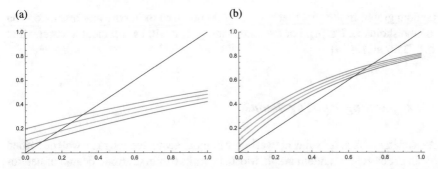

Fig. 5.2 Plot of f_ε for $\varepsilon \in \{0.05, 0.1, 0.15, 0.2\}$ and $c = 0.5$ (**a**), respectively, $c = 1.5$ (**b**). In black, the diagonal $h(z) = z$

On the other hand, for the case that $c > 1$ the fixed point of f_ε is lower bounded for all $\varepsilon > 0$ which is in line with Thm. 5.2, see Fig. 5.2b.

r ≥ 2: Also in this case the same heuristic reasoning as for $r = 1$ shows that

$$\alpha(\varepsilon) = \varepsilon + (1 - \varepsilon)\mathbb{P}(\mathrm{Poi}(c\alpha(\varepsilon)) \geq r)$$

for the fraction $\alpha(\varepsilon)$ of eventually infected nodes. This time, however, it is possible in general that the function

$$f_\varepsilon(z) := \varepsilon + (1 - \varepsilon)\mathbb{P}(\mathrm{Poi}(cz) \geq r)$$

has one, two, or three different fixed points in $(0, 1]$, depending on the values of c and ε. We can still describe the final infection fraction by choosing the smallest fixed point, but we require an additional condition: A fixed point $\hat{z}$ of f_ε is called *stable* if $f'_\varepsilon(\hat{z}) < 1$. The following result is then a special case of Janson et al. (2012) [Thm. 5.2.]:

Theorem 5.5 *Let $\hat{z}$ be the smallest fixed point of $f_\varepsilon(z)$ in $(0, 1]$ and assume that it is stable. Then, the fraction of eventually infected nodes converges to $\alpha(\varepsilon) = \hat{z}$ in probability.*

The theorem gives us a way to compute the final infection fraction for any given c and ε. In order to derive a statement about resilience of the network, note that $\mathbb{P}(\mathrm{Poi}(cz) \geq r) \leq \mathbb{P}(\mathrm{Poi}(cz) \geq 2) \leq (cz)^2/2$ and hence $f_\varepsilon(z) \leq \varepsilon + (cz)^2/2$. Thus the smallest fixed point $\hat{z}$ of f_ε is upper bounded by $(1 - \sqrt{1 - 2\varepsilon c^2})c^{-2}$ which tends to 0 as $\varepsilon \to 0$. Regardless of c we then obtain the following statement.

Theorem 5.6 *Consider the random graph model $G_{n,p}$ with $p = c/n$. If $r \geq 2$ (no contagious links), then the system is resilient.*

After having introduced our measure of systemic risk, we can now employ the resilience criteria formulated in Thms. 5.4 and 5.6 to derive the following management rules for the network thresholds to control systemic risk in the homogeneous random graph: In the case that $c < 1$, we do not need to impose any restrictions on the thresholds τ_i, $i \in [n]$. For the case that $c \geq 1$, it will be sufficient to require that $\tau_i \geq 2$ for all $i \in [n]$.

5.2.2 Getting Heterogeneous

As a matter of fact, only few networks are homogeneous enough to be well described by an Erdös-Rényi random graph. Most networks exhibit a strong degree of heterogeneity. The aim in this section is to describe an enhanced random graph model that overcomes this issue. Further, we change from the undirected random graph $G_{n,p}$ to a directed one since many real-world networks such as the network of interbank lending are directed. The model we present here was proposed in Detering et al. (2015) and is a directed version of the Chung-Lu inhomogeneous random graph (Chung and Lu 2002, 2003). The results presented in this section are special cases of results in Detering et al. (2015, 2016). Notable earlier works on the contagion process rBP in an undirected inhomogeneous random graph can be found in Amini et al. (2014) and Amini and Fountoulakis (2014).

We begin with a detailed description of the random graph model. We assign to each node $i \in [n]$ two weights: an *in-weight* w_i^- and an *out-weight* w_i^+. The in-weight describes the tendency of i to develop incoming edges (that is, edges pointing toward i), whereas the out-weight describes the tendency of developing outgoing edges (that is, edges pointing away from i). To formalize this, define for each possible edge $e = (i, j)$ going from node $i \in [n]$ to $i \neq j \in [n]$ the edge probability p_e by

$$p_e := \min\{1, n^{-1} w_i^+ w_j^-\}. \tag{5.2}$$

We denote the resulting random graph by $G_n(\mathbf{w}^-, \mathbf{w}^+)$, where $\mathbf{w}^- := (w_1^-, \ldots, w_n^-)$ and $\mathbf{w}^+ := (w_1^+, \ldots, w_n^+)$. The heterogeneity of the graph stems from assigning different weights to different nodes. In order to make statements about the graph in the limit $n \to \infty$, it is required that the graph grows in a somewhat regular fashion. In fact, we require that the fraction of nodes with weight level in any given interval stabilizes. To make this more precise, define the empirical distribution function

$$F_n(x, y) := n^{-1} \sum_{i \in [n]} \mathbf{1}\{w_i^- \leq x, w_i^+ \leq y\}.$$

and let (W_n^-, W_n^+) be a random vector distributed according to F_n. We shall assume that (W_n^-, W_n^+) converges in distribution to some random vector (W^-, W^+), and that furthermore $\mathbb{E}[W_n^-] \to \mathbb{E}[W^-] =: \lambda^- < \infty$ and $\mathbb{E}[W_n^+] \to \mathbb{E}[W^+] =: \lambda^+ < \infty$.

The random vector (W^-, W^+) serves as a limit object that is strongly associated with the sequence of random graphs $G_n(\mathbf{w}^-, \mathbf{w}^+)$ for $n \in \mathbb{N}$. We will see that it fully determines the degrees of its nodes and the outcome of the contagion process. As for the homogeneous random graph $G_{n,p}$, also in the heterogeneous setting we can describe the degree of each node in the limit $n \to \infty$. This time, every node $i \in [n]$ has an in-degree $\deg^-(i)$ and an out-degree $\deg^+(i)$. As in the homogeneous setting, their distribution is based on a Poisson distribution but also the weights w_i^- and w_i^+ play a role. More precisely, for large network sizes n it holds that $\deg^-(i) \sim \mathrm{Poi}(w_i^- \lambda^+)$ and $\deg^+(i) \sim \mathrm{Poi}(w_i^+ \lambda^-)$. To reverse the logic, it can be shown that in- and out-degree of each node function as maximum-likelihood estimators of its in- and out-weight (up to normalizing factors) when we want to calibrate our model parameters to some observed network structure. One can thus basically think of the weights in our model as the realized degrees of each node. It is hence no surprise that also the whole degree sequence is intimately related to the weight distribution. Consider the (random) empirical degree distribution

$$\tilde{F}_n(k, l) = n^{-1} \sum_{i \in [n]} \mathbf{1}\{\deg^-(i) \leq k, \deg^+(i) \leq l\}.$$

For a two-dimensional random vector (X, Y) let $Z = (\mathrm{Poi}(X), \mathrm{Poi}(Y))$ denote a two-dimensional mixed Poisson random vector with probability mass function given by

$$\mathbb{P}\left(Z = \binom{k}{j}\right) = \mathbb{E}\left[e^{-(X+Y)} \frac{X^k Y^j}{k! \, j!}\right].$$

Then, the degrees in the network are described as follows:

Lemma 5.2 *The (random) empirical in- and out-degree distributions over all nodes converge to the distribution of the random vector* $(\mathrm{Poi}(W^- \lambda^+), \mathrm{Poi}(W^+ \lambda^-))$.

In particular, $G_n(\mathbf{w}^-, \mathbf{w}^+)$ has much more flexibility in its degree distribution than $G_{n,c/n}$. By choosing weights W^- and W^+ with infinite variance, it is even possible to describe networks whose degree distributions have unbounded second moment—a feature that is often observed in real networks. See Fig. 5.3 for an illustration of the

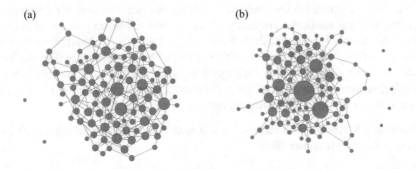

(a)　　　　　　　　　　(b)

Fig. 5.3 Typical configurations for $G_n(\mathbf{w}^-, \mathbf{w}^+)$ with $n = 100$ and Pareto-distributed weights with shape parameter **a** 3.5 (bounded second moment), respectively, **b** 2.5 (unbounded second moment). For simplicity, the graphs are depicted undirected. Node sizes scale with the corresponding degree

heterogeneity of $G_n(\mathbf{w}^-, \mathbf{w}^+)$. Also compare with Fig. 5.1. All three figures show graphs with exactly 200 edges but differ in the realized degree sequences due to the different choices for the weight distributions.

For the description of the contagion process, note that in real networks not only the network topology is very heterogeneous, but also the strengths of the different nodes. In the previous section, we described the contagion process by $r\mathsf{BP}$. However, there might be nodes that can endure more defaults of their neighbors than others. Therefore, we assign an *individual* threshold $\tau_i \in \mathbb{N} \cup \{\infty\}$ to each node describing the number of neighbors of i that need to become infected before i becomes infected as well. For example, in a banking network, τ_i can be thought of as the capital of some bank i. Let similarly as before (W_n^-, W_n^+, T_n) be a random variable with distribution equal to the empirical distribution of the weights and thresholds and assume that also for the extended network, (W_n^-, W_n^+, T_n) converges in distribution to some random vector (W^-, W^+, T). Similar as in the previous section, under these mild assumptions it is then possible to determine the fraction of eventually infected nodes by computing the smallest fixed point of a certain function. For $\varepsilon > 0$ let

$$f_\varepsilon(z) := \varepsilon \mathbb{E}[W^+] + (1 - \varepsilon)\mathbb{E}[W^+ \mathbb{P}(\mathrm{Poi}(W^- z) \geq T)],$$

which is clearly a continuous function. Since $f_\varepsilon(0) = \varepsilon$ and $f_\varepsilon(\mathbb{E}[W^+]) \leq \mathbb{E}[W^+]$, there will always exist at least one fixed point $\hat{z}$ of f_ε within $(0, \mathbb{E}[W^+]]$. As before we call such a fixed point *stable* if f_ε is continuously differentiable at $\hat{z}$ with $f_\varepsilon'(\hat{z}) < 1$. Then, the following holds:

Theorem 5.7 *Let $\hat{z}$ be the smallest fixed point of f_ε and assume that it is stable. Then, the fraction of eventually infected nodes converges in probability to*

$$\alpha(\varepsilon) = \varepsilon + (1 - \varepsilon)\mathbb{E}[\mathbb{P}(\mathrm{Poi}(W^-\hat{z}) \geq T)].$$

For the sake of readability, we restrict ourselves to the typical case that $\hat{z}$ is stable in Thm. 5.7. For more general results, see Detering et al. (2015, 2016).

Having quantified the final infection fraction, we can then turn our attention to investigating the resilience properties of the generalized heterogeneous networks. These are intimately related to the behavior of $f_0(z)$ near $z = 0$. Assume that there is some $z_0 > 0$ such that $f_0(z) > z$ for all $z \in (0, z_0)$. Then, for each $\varepsilon > 0$ the smallest fixed point $\hat{z}$ of $f_\varepsilon(z)$ will always be larger than z_0 (see Fig. 5.4a) and the final fraction of infected nodes in the graph will w. h. p. be larger than $\mathbb{E}[\mathbb{P}(\mathrm{Poi}(W^- z_0) \geq T)])$. In particular, we derive the following theorem:

Theorem 5.8 *Assume that there is $z_0 > 0$ such that $f_0(z) > z$ for all $z \in (0, z_0)$. Then, the system is non-resilient.*

The assumption of this theorem is satisfied in particular if f_0 has right derivative larger than 1 at $z = 0$. The remaining cases, see also Fig. 5.4b for an illustration, are covered by the following result:

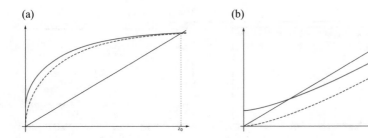

Fig. 5.4 Plot of the functions f_0 (dashed) and f_ε (solid) for $\varepsilon > 0$, $T = 2$ and weights $W^- = W^+$ Pareto distributed with shape parameter **a** 2.5 and **b** 3.5

Theorem 5.9 *Assume that $f_0(z)$ is continuously differentiable from the right at $z = 0$ with derivative $f_0'(0) < 1$. Then, the system is resilient.*

We again refer to Detering et al. (2015) for a version of Thm. 5.9 which makes weaker but also more technical assumptions on f_0.

Let us now discuss some consequences of Thms. 5.8 and 5.9 in more detail. Let us make the technical assumption $\mathbb{E}[W^-W^+] < \infty$. This is of course always satisfied if W^-, W^+ are independent, but it also captures many other cases in which there are significant correlations between the in- and out-degrees of the nodes: A characteristic setting is, for example, when $W^- \approx W^+$, and then the condition guarantees that both W^-, W^+ have bounded second moment. Under the assumption $\mathbb{E}[W^-W^+] < \infty$ we get the explicit representation

$$f_0'(z) = \mathbb{E}[W^-W^+\mathbb{P}(\text{Poi}(W^-z) = T - 1)]$$

and this is continuous for $z \in [0, \infty)$. Hence, Thms. 5.8 and 5.9 almost entirely characterize resilience in this case. In particular, if $T \geq 2$ almost surely, we get $f_0'(0) = 0$, and hence by Thm. 5.9, such networks are always resilient, which is consistent with our findings in the previous section. This readily yields sufficient requirements to make the system resilient.

However, note that the condition $\mathbb{E}[W^-W^+] < \infty$ is typically not satisfied for weights (i.e. degrees) with unbounded second moment which are frequently observed for real networks, such as interbank networks. It is then the case that also networks with $T \geq 2$ (or also $T \geq r$ for any $r \in \mathbb{N}$) almost surely can satisfy the condition in Thm. 5.8 and are hence non-resilient. See Fig. 5.5a for simulations on networks of finite size with weights $W^- = W^+$ according to a Pareto distribution with shape parameter 2.5 (i.e. with finite first moment but infinite second moment) and constant threshold $T = 2$. The final fraction of infected nodes concentrates around the asymptotic lower bound of 92.7% and the networks are hence non-resilient (with exception of only a few networks of very small size).

While the non-resilience property might be a favorable one in some applications where a large coverage of the network is targeted, for many others, such as financial networks, resilience is the desirable property. A characterization of resilient and

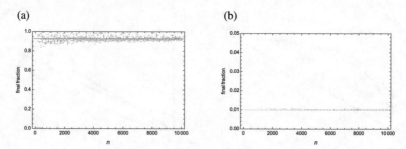

Fig. 5.5 Scatter plot of the final infection fraction for 10^4 simulations of finite networks. Weights $w_i^- = w_i^+$ are drawn from a Pareto distribution with shape parameter 2.5 and initially 1% of all nodes are infected. The thresholds are given by **a** $\tau_i = 2$, respectively, **b** $\tau_i = \max\{2, \alpha_c(w_i^-)^{\gamma_c}\}$. In **a**, the red line marks the asymptotic lower bound on the final infection fraction of about 92.7%

non-resilient networks also in the case that $\mathbb{E}[W^- W^+] = \infty$ is hence of high interest. The reason why $T \geq 2$ is not sufficient for resilience anymore is that there exist very strongly connected nodes in the network which either receive a lot of edges (have high in-weight) and are hence very susceptible or send a lot of edges (have high out-weight) and hence infect a large proportion of the network once they become infected. Typically in networks there are nodes which have both high in-weight and high out-weight which further increases the importance of their role in the infection process. Exactly these nodes are the ones that have to be equipped with higher thresholds when it comes to controlling systemic risk in the network.

Typically, as, for example, in the regulation of the financial sector, risk management strategies intend to ensure the survival of some given node by focusing on the risk exposures toward (i. e. infections from) other nodes and hence incoming links in the network. In this spirit, we aim to state threshold requirements which depend on a node's in-weight. More precisely, we intend to characterize resilience/non-resilience properties for heterogeneous networks where the thresholds τ_i are given by $\tau(w_i^-)$ for some non-decreasing integer-valued function $\tau : \mathbb{R} \to \mathbb{N}$. This then allows for the management of systemic risk by deriving sufficient threshold requirements for each particular node simply by observing (estimating) the respective in-weight and plugging it into function τ.

We will see that the threshold requirements strongly depend on the tail of the weight distributions. What is typically observed for degree (weight) distributions of real networks is that they closely resemble Pareto distributions in their tail. We thus assume in the following that the weight distributions are Pareto distributions and refer to Detering et al. (2016) for results on more general weight distributions. That is, there exist parameters $\beta^-, \beta^+ > 2$ (in order to ensure integrability of W^- and W^+) and minimal weights $w_{\min}^-$ and $w_{\min}^+$ such that the weight densities are given by

$$f_{W^{\pm}} = (\beta^{\pm} - 1)(w_{\min}^{\pm})^{\beta^{\pm}-1} w^{-\beta^{\pm}} \mathbf{1}\{w \geq w_{\min}^{\pm}\}.$$

It will turn out that the quantities

$$\gamma_c := 2 + \frac{\beta^- - 1}{\beta^+ - 1} - \beta^- \quad \text{and} \quad \alpha_c := \frac{\beta^+ - 1}{\beta^+ - 2} w_{min}^+ (w_{min}^-)^{1-\gamma_c}$$

play a central role in determining sufficient threshold requirements. It holds that $\gamma_c < 0$ only if $\mathbb{E}[W^- W^+] < \infty$. In this case, it is hence sufficient to require $\tau_i \geq 2$ for all $i \in [n]$ as has been discussed above. For all other cases, we investigate systems with $\tau_i \approx \alpha(w_i^-)^\gamma$ for certain constants α and γ.

Theorem 5.10 *Let the weights W^- and W^+ be Pareto distributed with parameters $\beta^-, \beta^+ > 2$ and $w_{min}^-, w_{min}^+ > 0$ and assume that $\tau_i = \tau(w_i^-)$ for some function τ: $\mathbb{R} \to \mathbb{N} \setminus \{1\}$. Then, the system is resilient, if one of the following holds:*

1. $\gamma_c < 0$,
2. $\gamma_c = 0$ and $\liminf_{w \to \infty} \tau(w) > \alpha_c + 1$,
3. $\gamma_c > 0$ and $\liminf_{w \to \infty} w^{-\gamma_c} \tau(w) > \alpha_c$.

Note that Thm. 5.10 only derives sufficient requirements to make the system resilient. In Detering et al. (2016), it is shown that these requirements are actually sharp in the sense that networks become non-resilient for thresholds $\tau_i = \tau(w_i^-)$ if τ satisfies $\limsup_{w \to \infty} w^{-\gamma_c} \tau(w) < \tilde{\alpha}_c$ for a certain $\tilde{\alpha}_c > 0$ which depends on the dependence structure between W^- and W^+. In the case that the weights are comonotone (nodes with larger in-weights also have larger out-weights and vice versa), $\tilde{\alpha}_c$ coincides with α_c from Thm. 5.10.

Moreover, Thm. 5.10 only ensures resilience in the limit $n \to \infty$ and $\varepsilon \to 0$. The derived threshold requirements are, however, also applicable to reasonably sized finite networks with positive initial infection probability. See, for example, Fig. 5.5b for simulations on networks with sizes in $[10^2, 10^4]$ and enforced threshold requirements $\tau_i = \max\{2, \alpha_c(w_i^-)^{\gamma_c}\}$ for all $i \in [n]$. The observed amplification is almost negligible. It is hence possible to implement risk management strategies based on Thm. 5.10 for real networks. Usually (if $\gamma_c > 0$) such strategies require larger (more connected) nodes to ensure higher resistance (threshold). However, since $\beta^-, \beta^+ > 2$, it holds that $\gamma_c < 1$ and the threshold function τ thus only needs to increase sublinearly with the weight. Finally, it is an appealing feature of our formula that for each node i the required threshold τ_i can be computed locally, i.e. only using information about its own edges once α_c and γ_c are known. This contrasts our risk management strategies from other approaches, where always knowledge about the entire system needed to be assumed.

5.2.3   A Weighted Contagion Process

In the previous sections, the contagion process was always based on counting the number of infected neighbors. In the first step, a node became infected as soon as any of its neighbors became infected. Later, we allowed for $r \geq 2$ neighbors to default before a certain node became infected, and finally, we assigned to each node $i \in [n]$ an individual threshold value τ_i. For many applications, however, the mere counting of infected neighbors is not enough, since rather the *strength* of the links to these infected neighbors is the determining quantity. For instance, in an interbank network, it is not the number of defaulted loans but rather their total amount that is relevant for the infection process. In this section, we therefore enhance our previous model once more to account for weighted edges.

In the specification of the random graph, we model the occurrence of edges as before by (5.2). Additionally, we assign to each node $j \in [n]$ a sequence of possible exposures $E_{1,j}, \ldots, E_{n,j}$ modeled by exchangeable $\mathbb{R}_+$-valued random variables, meaning that the order of the exposures does not influence their joint distribution. The random variable $E_{i,j}$ shall then describe a possible exposure from node i to node j. That is, we want to place it on an edge going from i to j if this edge is present in the graph. The assumption that the exposure list consists of exchangeable random variables is sensible for networks in which the strength of a link is determined by the receiving edge rather than by the sending edge (note that the exposure lists can significantly vary between different nodes $j \in [n]$).

In order to describe the contagion mechanism, we now assign to each node $i \in [n]$ an $\mathbb{R}_+$-valued parameter c_i resembling the strength of i. Motivated by the application to financial networks we will call c_i the capital of i hereafter. Similarly as previously we then describe the contagion process in the network as follows: At the beginning, a fraction $\varepsilon > 0$ of all nodes is infected. Other nodes in the network become infected as soon as their total exposure to infected nodes exceeds their capital. Note that our previous model is incorporated in this new model simply by choosing integer-valued capitals and $E_{i,j} = 1$ for all $i \neq j$. In this case, the capitals c_i had the interpretation of threshold values. In analogy to the previous model, we therefore introduce for each node $i \in [n]$ a threshold value τ_i which shall count the number of neighbors that can cause the infection of i. To be more precise, τ_i shall be the smallest integer value such that $\sum_{\ell \leq \tau_i} E_{\ell,i} \geq c_i$ if such a value exists. If $\sum_{\ell=1}^{n} E_{\ell,i} < c_i$, we simply set $\tau_i = \infty$. Then, τ_i is a random variable and it only describes a *hypothetical* threshold value since usually the nodes will not become infected in their natural order during the infection process. We now assume that still in the limit when the network size $n \to \infty$ the thresholds are described by a random variable T. Then, due to exchangeability of the exposure random variables and large network effect, we can restate Thm. 5.7 for this new model, where again

$$f_\varepsilon(z) := \varepsilon \mathbb{E}[W^+] + (1 - \varepsilon)\mathbb{E}[W^+ \mathbb{P}(\text{Poi}(W^- \hat{z}) \geq T)].$$

Theorem 5.11 *Let $\hat{z}$ be the smallest fixed point of f_ε and assume that it is stable. Then, the fraction of eventually infected nodes converges in probability to*

$$\alpha(\varepsilon) = \varepsilon + (1 - \varepsilon)\mathbb{E}[\mathbb{P}(\text{Poi}(W^- \hat{z}) \geq T)].$$

Also the results about non-resilience and resilience of the network generalize to the more complex setting.

Theorem 5.12 *Assume that there is $z_0 > 0$ such that $f_0(z) > z$ for all $z \in (0, z_0)$. Then, the system is non-resilient.*

Theorem 5.13 *Assume that $f_0(z)$ is continuously differentiable from the right at $z = 0$ with derivative $f_0'(0) < 1$. Then, the network is resilient.*

Finally, under some rather mild assumptions on the exposure sequences such as $\mathbb{E}[E_{j,i}] = \mu_i$ for all $j \in [n]$, we can also reformulate Thm. 5.10 for the new model which equips us with a formula for sufficient capital requirements to secure a system. See Detering et al. (2016) for a precise formulation of Thm. 5.14 and its assumptions.

Theorem 5.14 *Let the weights W^- and W^+ be Pareto distributed with parameters $\beta^-, \beta^+ > 2$ and $w_{min}^-, w_{min}^+ > 0$. Further assume that $c_i > \max_{j \in [n]} E_{j,i}$ (the capital is larger than the largest exposure) almost surely for all $i \in [n]$. Then, the following holds:*

1. *If $\gamma_c < 0$, then the system is resilient.*

If additionally, there exists a function $\tau : \mathbb{R} \to \mathbb{N}$ such that the capitals satisfy $c_i \geq \tau(w_i^-)\mu_i$ almost surely for all $i \in [n]$, then the system is resilient if one of the following holds:

1. *$\gamma_c = 0$ and $\liminf_{w \to \infty} w^{-\gamma} \tau(w) > 0$ for some $\gamma > 0$,*
2. *$\gamma_c > 0$ and $\liminf_{w \to \infty} w^{-\gamma_c} \tau(w) > \alpha_c$.*

In particular, in the usual case that $\gamma_c > 0$ (e.g., $\beta^-, \beta^+ < 3$) Thm. 5.14 ensures resilience if each institution i holds capital larger than $\alpha_c(w_i^-)^{\gamma_c}\mu_i$ (and not less than its largest exposure). As before, this is a quantity that can be computed by each institution individually simply by counting and averaging their exposures in the

(a)   (b)

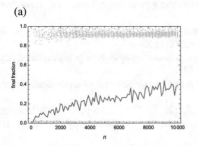
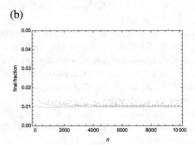

Fig. 5.6 Scatter plot of the final infection fraction for 10^4 simulations of finite weighted networks. Node-weights $w_i^- = w_i^+$ are drawn from a Pareto distribution with shape parameter 2.5 as are the edge-weights $E_{j,i}$, and initially 1% of all nodes are infected. The capitals are given by **a** $c_i = 1.001 \max_{j \in [n]} E_{j,i}$, respectively, **b** $c_i = \max\{1.001 \max_{j \in [n]} E_{j,i}, \alpha_c(w_i^-)^{\gamma_c}\mu_i\}$. In **a**, the red line marks the average final fraction over all 100 simulations for each network size

system. The theorem hence provides us with an easy applicable risk management policy to prevent networks from systemic risk. To test its applicability, we pursue simulations similar to the ones from Figs. 5.5a and b but enrich the network with Pareto-distributed exposures with shape parameter 2.5. As can be seen from Fig. 5.6a it is not sufficient to only prohibit contagious links in the network in order to make the system resilient. The derived capital requirements from Thm. 5.14 on the other hand ensure resilience of the system as can be seen from Fig. 5.6b. Note that the outcome of the simulation is more volatile than for the threshold model from Subsect. 5.2.2 since also the exposure sizes carry a lot of randomness. Still our derived capital requirements work very well to contain the infection.

References

Alon, N. & Spencer, J. H. (2016), *The Probabilistic Method*, 4th edn, Wiley Publishing.

Amini, H. & Fountoulakis, N. (2014), 'Bootstrap percolation in power-law random graphs', *Journal of Statistical Physics* **155**(1), 72–92. http://dx.doi.org/10.1007/s10955-014-0946-6

Amini, H., Filipović, D. & Minca, A. (2015), 'Systemic risk and central clearing counterparty design', *Swiss Finance Institute Research Paper* pp. 13–34.

Amini, H., Fountoulakis, N. & Panagiotou, K. (2014), 'Bootstrap percolation in inhomogeneous random graphs', *arXiv:1402.2815*

Amini, H., Cont, R. & Minca, A. (2016), 'Resilience to contagion in financial networks', *Mathematical Finance* **26**(2), 329–365.

Armenti, Y., Crépey, S., Drapeau, S. & Papapantoleon, A. (2018), 'Multivariate shortfall risk allocation and systemic risk', *SIAM Journal on Financial Mathematics* **9**(1), 90–126.

Artzner, P., Delbaen, F., Eber, J. & Heath, D. (1999), 'Coherent measures of risk', *Mathematical Finance* **9**(3), 203–228.

Biagini, F., Fouque, J.-P., Fritelli, M. & Meyer-Brandis, T. (2019+), 'A unified approach to systemic risk measures via acceptance sets', *Mathematical Finance* .

Bollobás, B. (2001), *Random Graphs*, Cambridge Studies in Advanced Mathematics, 2nd edn, Cambridge University Press, Cambridge.

Chalupa, J., Leath, P. L. & Reich, G. R. (1979), 'Bootstrap percolation on a Bethe lattice', *Journal of Physics C: Solid State Physics* **12**(1), L31–L35.

Chen, C., Iyengar, G. & Moallemi, C. (2013), 'An axiomatic approach to systemic risk', *Management Science* **59**(6), 1373–1388.

Chong, C. & Klüppelberg, C. (2018), 'Contagion in financial systems: A Bayesian network approach', *SIAM Journal on Financial Mathematics* **9**(1), 28–53.

Chung, F. & Lu, L. (2002), 'Connected components in random graphs with given expected degree sequences', *Annals of Combinatorics* **6**(2), 125–145.

Chung, F. & Lu, L. (2003), 'The average distances in random graphs with given expected degrees', *Internet Mathematics* **1**, 91–114.

Detering, N., Meyer-Brandis, T. & Panagiotou, K. (2015), 'Bootstrap percolation in directed and inhomogeneous random graphs', *Electronic Journal of Combinatorics,* **26**(2), 2019, arXiv:1511.07993.

Detering, N., Meyer-Brandis, T., Panagiotou, K. & Ritter, D. (2016), 'Managing default contagion in inhomogeneous financial networks', *SIAM Journal on Financial Mathematics,* **10**(2), 2019, arXiv:1610.09542.

Eisenberg, L. & Noe, T. H. (2001), 'Systemic risk in financial systems', *Management Science* **47**(2), 236–249.

Erdős, P. & Rényi, A. (1960), On the evolution of random graphs, *in* 'Publication of the Mathematical Institute of the Hungarian Academy of Sciences', pp. 17–61.

Feinstein, Z., Rudloff, B. & Weber, S. (2017), 'Measures of systemic risk', *SIAM Journal on Financial Mathematics* **8**(1), 672–708.

Fouque, J.-P. & Langsam, J. A., eds (2013), *Handbook on systemic risk*, Cambridge University Press, Cambridge.

Gai, P. & Kapadia, S. (2010), 'Contagion in Financial Networks', *Proceedings of the Royal Society A* **466**, 2401–2423.

Gandy, A. & Veraart, L. A. M. (2017), 'A Bayesian methodology for systemic risk assessment in financial networks', *Management Science* **63**(12), 4428–4446.

Gilbert, E. N. (1959), 'Random Graphs', *Ann. Math. Statist.* **30**(4), 1141–1144. https://doi.org/10.1214/aoms/1177706098

Hoffmann, H., Meyer-Brandis, T. & Svindland, G. (2016), 'Risk-consistent conditional systemic risk measures', *Stochastic Processes and their Applications* **126**(7), 2014–2037.

Hoffmann, H., Meyer-Brandis, T. & Svindland, G. (2018), 'Strongly consistent multivariate conditional risk measures', *Mathematics and Financial Economics* **12**(3), 413–444.

Hofstad, R. v. d. (2016), *Random Graphs and Complex Networks*, Vol. 1 of *Cambridge Series in Statistical and Probabilistic Mathematics*, Cambridge University Press.

Hurd, T. R. (2016), *Contagion! Systemic Risk in Financial Networks*, Springer.

Janson, S., Łuczak, T. & Rucinski, A. (2011), *Random Graphs*, Wiley.

Janson, S., Łuczak, T., Turova, T. & Vallier, T. (2012), 'Bootstrap percolation on the random graph $G_{n,p}$', *Ann. Appl. Probab.* **22**(5), 1989–2047. http://dx.doi.org/10.1214/11-AAP822

Kromer, E., Overbeck, L. & Zilch, K. (2016), 'Systemic risk measures on general measurable spaces', *Mathematical Methods of Operations Research* **84**(2), 323–357.

Kusnetsov, M. & Veraart, L. A. M. (2016), 'Interbank Clearing in Financial Networks with Multiple Maturities', *Preprint* .

Weber, S. & Weske, S. (2017), 'The joint impact of bankruptcy costs, cross-holdings and fire sales on systemic risk in financial networks', *Probability, Uncertainty and Quantitative Risk* **2**(9), 1–38.

Chapter 6
Bayesian Networks for Max-Linear Models

Claudia Klüppelberg and Steffen Lauritzen

Abstract We study Bayesian networks based on max-linear structural equations as introduced in Gissibl and Klüppelberg (2018) and provide a summary of their independence properties. In particular, we emphasize that distributions for such networks are generally not faithful to the independence model determined by their associated directed acyclic graph. In addition, we consider some of the basic issues of estimation and discuss generalized maximum likelihood estimation of the coefficients, using the concept of a generalized likelihood ratio for non-dominated families as introduced by Kiefer and Wolfowitz (1956). Finally, we argue that the structure of a minimal network asymptotically can be identified completely from observational data.

6.1 Introduction

The type of model we are studying has been motivated by applications to risk analysis, where extreme risks play an essential role and may propagate through a network. For example, say, if an extreme rainfall happens on a specific location near a river network, it may effect water levels at other parts of the network in an essentially deterministic fashion. Similar phenomena occur in the analysis of risk for other complex systems.

Specifically, the model presented in (6.1) below arose in the context of technical risk analysis, more precisely, in an investigation of the "runway overrun" event of airplane landing. Numerous variables contribute to this event and extraordinary values of some variables lead invariably to a runway overrun (see Gissibl et al. 2017 for more details) naturally leading to questions about cause and effect of risky events. Other potential examples for risk-related cause and effect relations include chemical

C. Klüppelberg (✉)
Center for Mathematical Sciences, Technical University of Munich, Munich, Germany
e-mail: cklu@ma.tum.de

S. Lauritzen
Department of Mathematical Sciences, University of Copenhagen, Copenhagen, Denmark
e-mail: lauritzen@math.ku.dk

© Springer Nature Switzerland AG 2019   79
F. Biagini et al. (eds.), *Network Science*, https://doi.org/10.1007/978-3-030-26814-5_6

pollution of rivers (Hoef et al. 2006), flooding in river networks (Asadi et al. 2015), financial risk (Einmahl et al. 2018), and many others.

Statistical theory and applications of extreme value theory until the 1990s mainly focused on i.i.d. data as, for instance, yearly maximal water levels to predict future floodings or peaks over thresholds used to estimate the value-at-risk (Embrechts et al. 1997). From this, both theory and applications moved on to multivariate data, modelling risks like joint wind and wave extremes as well as extreme risks in financial portfolios (Beirlant et al. 2006). The investigation of extremes in time series models has proved useful in financial and environmental risk analysis, and also in telecommunication (Finkenstädt and Rootzén 2004). More recently, extreme space-time models have been suggested and applied to environmental risk data (Buhl et al. 2016; Davis et al. 2013; Davison et al. 2012; Huser and Davison 2014).

The paper focuses on first steps reporting on the methodological development associated with a specific class of network models. We begin with introducing our leading example of a recursive max-linear model which is Example 2.1 of Gissibl and Klüppelberg (2018):

Example 6.1 Consider the network in the figure below:

Each node i in the network represents a random variable X_i, and the joint distribution of $X = (X_1, X_2, X_3, X_4)$ is determined by a system of *max-linear structural equations*

$$X_1 = Z_1, \quad X_2 = \max(c_{21}X_1, Z_2), \quad X_3 = \max(c_{31}X_1, Z_3),$$
$$X_4 = \max(c_{42}X_2, c_{43}X_3, Z_4),$$

where Z_1, Z_2, Z_3, Z_4 are independent positive random variables and the coefficients c_{ji} are all strictly positive.

The interpretation of a system like this is that each node in the network is subjected to a random shock Z_i and the effect from shocks of other nodes pointing to it, the latter being attenuated or amplified by the coefficients c_{ji}. To simplify notation here and later we write $a \vee b$ for $\max(a, b)$. We can alternatively represent $X = (X_1, X_2, X_3, X_4)$ directly in terms of the noise variables as

$$X_1 = Z_1$$
$$X_2 = c_{21}X_1 \vee Z_2 = c_{21}Z_1 \vee Z_2$$
$$X_3 = c_{31}X_1 \vee Z_3 = c_{31}Z_1 \vee Z_3$$

$$X_4 = c_{42}X_2 \vee c_{43}X_3 \vee Z_4$$
$$= c_{42}(c_{21}Z_1 \vee Z_2) \vee c_{43}(c_{31}Z_1 \vee Z_3) \vee Z_4$$
$$= (c_{42}c_{21} \vee c_{43}c_{31})Z_1 \vee c_{42}Z_2 \vee c_{43}Z_3 \vee Z_4.$$

We may then summarize the above coefficients to the noise variables $Z_1, \ldots, Z_4$ in the matrix

$$B = \begin{pmatrix} 1 & 0 & 0 & 0 \\ c_{21} & 1 & 0 & 0 \\ c_{31} & 0 & 1 & 0 \\ c_{42}c_{21} \vee c_{43}c_{31} & c_{42} & c_{43} & 1 \end{pmatrix},$$

In greater generality, we may write such a *recursive max-linear model* as

$$X_v = \bigvee_{u \in pa(v)} c_{vu}X_k \vee c_{vv}Z_v, \quad v = 1, \ldots, d, \tag{6.1}$$

where $pa(v)$ denotes parents of v in a directed acyclic graph (DAG) and Z_v represent independent noise variables. The present article is concerned with such models and summarizes basic elements of Gissibl and Klüppelberg (2018) and Gissibl et al. (2019).

In this setting, natural candidates for the noise distributions are extreme value distributions or distributions in their domains of attraction resulting in a corresponding multivariate distribution with dependence structure given by the DAG (for details and background on multivariate extreme value models, see, e.g., Haan and Ferreira 2006; Resnick 1987, 2007).

The paper is structured as follows. In Sect. 6.2, we establish the necessary terminology (Subsect. 6.2.1) and introduce Bayesian networks (Subsect. 6.2.2), and basic properties of conditional independence (Subsect. 6.2.3). In Subsect. 6.2.4 we establish basic Markov properties of Bayesian networks. In Sect. 6.3, we study the specific Markov properties of Bayesian networks given by max-linear structural equations as in (6.1) and in Sect. 6.4 we study statistical properties of the models.

6.2 Preliminaries

6.2.1 Graph Terminology

A *graph* as we use it here is determined by a finite vertex set V, an edge set E, and a map that to each edge e in E associates its endpoints $u, v \in V$. Our graphs are *simple* so that there are no self-loops (edges with identical endpoints) and no multiple edges. Therefore, we can identify an edge e with its endpoints u, v so we can write $e = uv$. An edge uv of a *directed* graph points *from u to v* and we write $u \to v$. Then, u is a

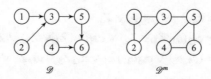

Fig. 6.1 A DAG $\mathscr{D}$ and its moral graph $\mathscr{D}^m$. In $\mathscr{D}$, 3 has parents 1, 2 and 5 is a child of 3. The DAG $\mathscr{D}$ is a polytree. The node 6 is a descendant of 1, and 2 is an ancestor of 5. The set $\{1, 2, 3, 5\}$ is ancestral in $\mathscr{D}$. With the node numbering given, the DAG is well-ordered

parent of v and v is a *child* of u. The set of parents of v is denoted pa(v) and the set of children of u is ch(u). If uv is an edge, we also say that u and v are adjacent and write $u \sim v$ whether or not the edge is directed.

A *walk* ω from u to v of *length n* is a sequence of vertices $\omega = [u = u_0, u_1, \ldots, u_n = v]$ so that $u_{i-1} \sim u_i$ for all $i = 1, \ldots, n$. A walk is a *cycle* if $u = v$. A *path* is a walk with no repeated vertices. The walk is *directed* from u to v if $u_{i-1} \to u_i$ for all i. If all edges in a graph $\mathscr{D} = (V, E)$ are directed, $\mathscr{D}$ is a *directed graph*. A directed graph is *acyclic* if it has no directed cycles. A *DAG* is a directed acyclic graph. A DAG is a *tree* if every vertex has at most one parent and a *polytree* if there is at most one path between two vertices u and v.

If there is a directed path from u to v in $\mathscr{D}$, we say that u is an *ancestor* of v and v a *descendant* of u and write $u \rightsquigarrow v$ or $v \leftsquigarrow u$. The set of ancestors of v is denoted an(v). A set $A \subseteq V$ is said to be *ancestral* if an(v) $\subset A$ for all $v \in A$, or, alternatively, if pa(v) $\subset A$ for all $v \in A$. For a subset A of V we let An(A) denote the smallest ancestral set containing A.

We say that the vertex set V of a DAG $\mathscr{D}$ is *well-ordered* if $V = \{1, \ldots, d\}$ and all edges in $\mathscr{D}$ point from low to high, i.e. if $ij \in E \implies i < j$. Then, the set of *predecessors* of a vertex i is pr(i) $= \{1, \ldots, i - 1\}$.

For a DAG $\mathscr{D}$, we define its *moral graph* $\mathscr{D}^m$ as the simple, undirected graph with the same vertex set but with u and v adjacent in $\mathscr{D}^m$ if and only if either $u \sim v$ in $\mathscr{D}$ or if u and v have a common child. For further general graph terminology, we refer the reader to West (2001) but some of the concepts above are illustrated in Fig. 6.1.

6.2.2 Bayesian Networks

A real-valued Bayesian network associated with a given DAG $\mathscr{D} = (V, E)$ is determined by specifying random variables $X = (X_v, v \in V)$ and the conditional distribution of each of these, given values of their parent variables; for example, as

$$P(X_v \le x \mid X_{\text{pa}(v)}) = F(x \mid x_{\text{pa}}(v)).$$

Because there are no directed cycles in $\mathscr{D}$, there is a unique joint distribution corresponding to this specification.

Alternatively, as in Example 6.1, we can specify these conditional distributions through *structural equations* which describe the conditional distribution of X_v conditionally on $X_{pa(v)} = x_{pa(v)}$ in a functional form. More precisely a system of equations of the form

$$X_v = g_v(X_{pa(v)}, Z_v), \quad v \in V, \tag{6.2}$$

where $(Z_v)_{v \in V}$ are independent noise variables and g_v suitable functions.

A system of structural equations as above is sometimes referred to as a *data generating mechanism*, interpreting each equation as a way of generating random variables with the desired conditional distribution.

An important instance of these models is *linear structural equation models* where the functions g_v are linear and hence

$$X_v = \sum_{u \in pa(v)} c_{vu} X_u + c_{vv} Z_v, \quad v \in V, \tag{6.3}$$

where c_{vu}, $u \in pa(v)$, c_{vv} are *structural coefficients*, see, for example, Bollen (1989). In general, a structural equation system need not be associated with a DAG, but if it is, the equation system is said to be *recursive*.

If the distributions of Z_v have heavy tails and all structural coefficients are non-negative, the sum tends to be dominated by the largest term:

$$\sum_{u \in pa(v)} c_{vu} X_u + c_{vv} Z_v \approx \bigvee_{u \in pa(v)} c_{vu} X_u \vee c_{vv} Z_v$$

and hence for such cases, the max-linear variant in (6.4) is described in more detail in Sect. 6.3 below.

6.2.3 Conditional Independence

The notion of conditional independence is at the heart of graphical models, including Bayesian networks. For three random variables (X, Y, Z), we say that X is conditionally independent of Y given Z if the conditional distribution of X given (Y, Z) does not depend on Y and we then write $X \perp\!\!\!\perp Y \mid Z$ or $X \perp\!\!\!\perp_P Y \mid Z$ if we wish to emphasize the dependence on the joint distribution P of (X, Y, Z).

The notion of conditional independence has a number of important properties, see, e.g., Dawid (1980) or Lauritzen (1996).

Proposition 6.1 *Let $(\Omega, \mathbb{F}, P)$ be a probability space and X, Y, Z, W random variables on Ω. Then, the following properties hold.*

(C1) *If $X \perp\!\!\!\perp Y \mid Z$, then $Y \perp\!\!\!\perp X \mid Z$ (symmetry);*
(C2) *If $X \perp\!\!\!\perp Y \mid Z$ and $W = \phi(Y)$, then $X \perp\!\!\!\perp W \mid Z$ (reduction);*

(C3) *If $X \perp\!\!\!\perp (Y, Z) \mid W$, then $X \perp\!\!\!\perp Y \mid (Z, W)$ (weak union);*
(C4) *If $X \perp\!\!\!\perp Z \mid Y$ and $X \perp\!\!\!\perp W \mid (Y, Z)$, then $X \perp\!\!\!\perp (Z, W) \mid Y$ (contraction);*

It is occasionally important to abstract the notion of conditional independence away from necessarily being concerned with probability measures. An (abstract) *independence model* $\perp_\sigma$ over V is a ternary relation over subsets of a finite set V. The independence model is a *semi-graphoid* if the following holds for mutually disjoint subsets A, B, C, D:

(S1) *If $A \perp_\sigma B \mid C$, then $B \perp_\sigma A \mid C$ (symmetry);*
(S2) *If $A \perp_\sigma (B \cup D) \mid C$, then $A \perp_\sigma B \mid C$ and $A \perp_\sigma D \mid C$ (decomposition);*
(S3) *If $A \perp_\sigma (B \cup D) \mid C$, then $A \perp_\sigma B \mid (C \cup D)$ (weak union);*
(S4) *If $A \perp_\sigma B \mid C$ and $A \perp_\sigma D \mid (B \cup C)$, then $A \perp_\sigma (B \cup D) \mid C$ (contraction);*

Further, the independence model is a *graphoid* if it also satisfies

(S5) *If $A \perp_\sigma B \mid (C \cup D)$ and $A \perp_\sigma C \mid (B \cup D)$, then $A \perp_\sigma (B \cup C) \mid D$ (intersection).*

We shall in particular be interested in distributions on product spaces $\mathcal{X} = \times_{v \in V} \mathcal{X}_v$ where V is a finite set. For $A \subseteq V$, we write $x_A = (x_v, v \in A)$ to denote a generic element in $\mathcal{X}_A = \times_{v \in A} \mathcal{X}_v$, and similarly $X_A = (X_v)_{v \in A}$.

If P is a probability distribution on $\mathcal{X}$, we can now define an independence model $\perp\!\!\!\perp$ by the relation

$$A \perp\!\!\!\perp B \mid C \iff X_A \perp\!\!\!\perp_P X_B \mid X_C$$

and it follows from Prop. 6.1 that $\perp\!\!\!\perp$ is a semi-graphoid; in general $\perp\!\!\!\perp$ is not a graphoid without further assumptions on P.

Another important independence model is determined by *separation* in an undirected graph. More precisely, if $\mathcal{G} = (V, E)$ is an undirected graph we can define an independence model $\perp_\mathcal{G}$ by letting $A \perp_\mathcal{G} B \mid S$ mean that all paths in $\mathcal{G}$ from A to B intersect S. Then, it is easy to see that $\perp_\mathcal{G}$ is always a graphoid; indeed the term graphoid refers to this fact.

For a directed graph, the relevant notion of separation is more subtle. A vertex u is a *collider* on a path π if two arrowheads meet on the walk at u, i.e. if the following situation occurs $\pi = [\cdots \to u \leftarrow \cdots]$.

We say that a path π from u to v in a DAG $\mathcal{D}$ is *connecting* relative to S, if all colliders on π are in the ancestral set $\text{An}(S)$, and all non-colliders are outside S. A path that is not connecting relative to S is said to be *blocked* by S. We then define an independence model $\perp_\mathcal{D}$ relative to a directed graph $\mathcal{D}$ as follows:

Definition 6.1 For three disjoint subsets A, B, and S of the vertex set V of a graph $\mathcal{G} = (V, E)$, we say that A and B are $\mathcal{D}$-separated by S if all paths from A to B are blocked by S and we then write $A \perp_\mathcal{D} B \mid S$.

Example 6.2 Consider the network in the figure below, only slightly more complex than in Example 6.1:

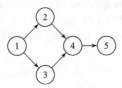

We have $2 \perp_{\mathscr{D}} 3 \mid 1$ since the path $2 \leftarrow 1 \rightarrow 3$ is blocked as the non-collider 1 is in $S = \{1\}$ whereas the path $2 \rightarrow 4 \rightarrow 3$ is blocked because the collider 4 is not an ancestor of $S = \{1\}$; on the other hand it holds that $\neg(2 \perp_{\mathscr{D}} 3 \mid \{1, 5\})$ since now the second path is rendered active as the collider 4 is in $\mathrm{An}(\{1, 5\})$.

Note that this definition in a natural way extends that of $\perp_{\mathscr{G}}$ for an undirected graph, as an undirected graph does not have colliders. The independence model $\perp_{\mathscr{D}}$ also satisfies the graphoid axioms, see, e.g., Lauritzen and Sadeghi (2018).

There is an alternative method for checking $\mathscr{D}$-separation in terms of standard separation in a suitable undirected graph, associated with the query. More precisely we say that A is m-separated from B by S and we write $A \perp_m B \mid S$ if S separates A from B in the moral graph $(\mathscr{D}_{\mathrm{An}(A \cup B \cup S)})^m$. We then have:

Proposition 6.2 *Let A, B and S be disjoint subsets of the nodes of a directed acyclic graph $\mathscr{G}$. Then, $A \perp_{\mathscr{D}} B \mid S \iff A \perp_m B \mid S$.*

For a proof, see Richardson (2003), amending an inaccuracy in Lauritzen et al. (1990).

Example 6.3 To illustrate the alternative procedure, we again consider the network in Example 6.2

If we wish to check whether $2 \perp_{\mathscr{D}} 3 \mid 1$ we consider the subgraph induced by the ancestral set of $\{1, 2, 3\}$ and moralize to obtain the graph to the left in the figure below. Since 1 is a separator in this graph, we conclude that $2 \perp_{\mathscr{D}} 3 \mid 1$.

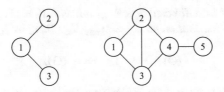

On the other hand, if the query is whether $2 \perp_{\mathscr{D}} 3 \mid \{1, 5\}$ we have $\mathrm{An}(\{1, 5\}) = V$ and thus the relevant moral graph is given to the right in the figure above; in this graph, 2 and 3 are not separated by $\{1, 5\}$ so we conclude $\neg(2 \perp_{\mathscr{D}} 3 \mid \{1, 5\})$.

6.2.4   Markov Properties of Bayesian Networks

It follows directly from the construction of a Bayesian network that the joint distribution P satisfies the *well-ordered Markov property* (O) w.r.t. $\mathcal{D}$ if for some well-ordering of V, every variable is conditionally independent of its predecessors given its parents

$$v \perp\!\!\!\perp \mathrm{pr}(v) \mid \mathrm{pa}(v)$$

for all $v \in V = \{1, \ldots, d\}$.

We further say that P obeys the *local Markov property* (L) w.r.t. $\mathcal{D}$ if every variable is conditionally independent of its non-descendants, given its parents:

$$v \perp\!\!\!\perp (\mathrm{nd}(v) \setminus \mathrm{pa}(v)) \mid \mathrm{pa}(v).$$

And, finally, P satisfies the *global Markov property* (G) w.r.t. $\mathcal{D}$ if

$$A \perp_{\mathscr{D}} B \mid C \implies A \perp\!\!\!\perp B \mid C.$$

Example 6.4 Consider the network in the figure below:

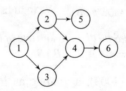

The numbering of the nodes here constitute a well-ordering so, for example, (O) implies $5 \perp\!\!\!\perp \{1, 3, 4\} \mid 2$, whereas the local Markov property (L) implies $5 \perp\!\!\!\perp \{1, 3, 4, 6\} \mid 2$; the global Markov property implies, for example, $5 \perp\!\!\!\perp \{1, 6\} \mid 4$.

In the case of undirected graphs, the local and global Markov properties are different (Lauritzen 1996, Sect. 3.2), but here we have

Theorem 6.1 *Let $\mathcal{D}$ be a directed acyclic graph with $V = \{1, \ldots, d\}$ well-ordered and P a probability distribution on $\mathscr{X} = \times_{v \in V} \mathscr{X}_v$. Then, we have*

$$(\mathrm{O}) \iff (\mathrm{L}) \iff (\mathrm{G}).$$

In words, if P satisfies any of these Markov properties, it satisfies all of them.

Proof This fact is established in Lauritzen et al. (1990) [Corollary 2] for any semi-graphoid independence model $\perp_\sigma$. □

Note that in particular it is true that *if P satisfies (O) w.r.t. one well-ordering, it satisfies (O) w.r.t. all well-orderings.*

The global Markov property gives a sufficient condition for conditional independence in terms of $\mathscr{D}$-separation. Another central concept is that of faithfulness, formally defined below

Definition 6.2 A probability distribution P on $\mathscr{X} = \times_{v \in V} \mathscr{X}_v$ is said to be *faithful* to a DAG $\mathscr{D}$ if

$$A \perp_{\mathscr{D}} B \mid C \iff A \perp\!\!\!\perp_P B \mid C.$$

In other words, if $\mathscr{D}$-separation is also necessary for conditional independence.

Generally, most probability distributions are faithful (Meek 1995), but we shall later see that this is not the case for the special Bayesian networks we study here.

Finally, we need to emphasize that two different DAGs can define exactly the same independence model. Consider two graphs $\mathscr{D}_1$ and $\mathscr{D}_2$ as well as their associated independence models $\perp_{\mathscr{D}_1}$ and $\perp_{\mathscr{D}_2}$. It may well happen that even though the graphs are different, their independence models might be identical, see, for example, Fig. 6.2 below.

Fig. 6.2 The three DAGs to the left of the figure are Markov equivalent; the only non-trivial element of their independence models is $u \perp_{\mathscr{D}} w \mid v$. The DAG to the right in the figure has a different independence model, since there $u \perp_{\mathscr{D}} w$

Here, all independence models are the same although the graphs are different. This also means that any probability distribution P which satisfies the global Markov property for any of them, automatically satisfies the global Markov property for all of them. We formally define

Definition 6.3 Two DAGs $\mathscr{D}_1$ and $\mathscr{D}_2$ are *Markov equivalent* if and only if their independence models coincide, i.e. if $A \perp_{\mathscr{D}_1} B \mid C \iff A \perp_{\mathscr{D}_2} B \mid C$.

The following result was shown by Frydenberg (1990) and Verma and Pearl (1990) and gives a necessary and sufficient condition for two DAGs to be Markov equivalent.

Theorem 6.2 *Two directed acyclic graphs $\mathscr{D}_1 = (V, E_1)$ and $\mathscr{D}_2 = (V, E_2)$ are Markov equivalent if and only if they have the same skeleton $\mathrm{ske}(\mathscr{D}_1) = \mathrm{ske}(\mathscr{D}_2)$ and the same unshielded colliders.*

Here, the *skeleton* $\mathrm{ske}(\mathscr{D})$ of a DAG $\mathscr{D}$ is the undirected graph with $u \sim v$ in $\mathrm{ske}(\mathscr{D})$ if $u \sim v$ in $\mathscr{D}$, and an *unshielded collider* is a triple $u \to w \leftarrow v$ with $u \nsim v$.

6.3   Recursive Max-Linear Structural Equation Models

We shall be interested in Bayesian networks defined through structural equation systems (6.2) where the functions g_v are *max-linear,* i.e. the additions in (6.3) are replaced with the operation of forming the maximum.

Henceforth, we assume that the vertex set of our DAG $\mathcal{D} = (V, E)$ is well-ordered so $V = \{1, \ldots, d\}$ and assume a data generating mechanism specified via a *recursive max-linear structural equation model*, which has representation

$$X_v = \bigvee_{u \in \mathrm{pa}(v)} c_{vu} X_u \vee c_{vv} Z_v, \quad v = 1, \ldots, d, \tag{6.4}$$

where $Z_1, \ldots, Z_d$ are independent and identically distributed with a continuous distribution having support $\mathbb{R}_+ = (0, \infty)$, and $c_{vu} > 0$, $u \in \mathrm{pa}(v)$, c_{vv} are *structural coefficients* in the equations or *edge weights* for the associated DAG $\mathcal{D}$.

Following Gissibl and Klüppelberg (2018) we say this is a *recursive max-linear model*. Note that our use of indices for edge weights here is the opposite of that used in Gissibl and Klüppelberg (2018).

For simplicity, we assume throughout the rest of the paper that $c_{vv} = 1$ for all $v \in V$. For a path $\pi = [u = k_0 \to k_1 \to \cdots \to k_n = v]$ of length n from u to v, we define the quantities

$$d_{vu}(\pi) := \prod_{l=0}^{n-1} c_{k_{l+1} k_l} \quad \text{and} \quad b_{vu} := \bigvee_{\pi \in \Pi_{uv}} d_{vu}(\pi), \tag{6.5}$$

where Π_{uv} denotes all paths from u to v. In summary, we define

$$b_{vu} = \bigvee_{\pi \in \Pi_{uv}} d_{vu}(\pi) \text{ for } u \in \mathrm{an}(v); \quad b_{vv} = c_{vv} = 1; \quad b_{vu} = 0 \text{ for } u \in V \setminus \mathrm{An}(v),$$

$$\tag{6.6}$$

where $\mathrm{An}(v) = \mathrm{an}(v) \cup \{v\}$ is the smallest ancestral set containing vertex v. We then arrange these coefficients in the *max-linear coefficient matrix* $B = (b_{vu})_{d \times d}$ and find

$$X_v = \bigvee_{u \in \mathrm{An}(v)} b_{vu} Z_u, \quad v = 1, \ldots, d. \tag{6.7}$$

This equation represents X as a *max-linear model* as defined for instance in Wang and Stoev (2011).

For two non-negative matrices F and G, where the number n of columns in F is equal to the number of rows in G, we introduce the product $\odot$ as

$$(F \odot G)_{vu} = \left(\bigvee_{k=1}^{n} f_{vk} g_{ku} \right). \tag{6.8}$$

If we collect the noise variables into the column vector $Z = (Z_1, \ldots, Z_d)'$, the representation (6.7) of X can then be written as

$$X = B \odot Z = \Big(\bigvee_{u=1}^{d} b_{vu} Z_j, i = 1, \ldots, d \Big) = \Big(\bigvee_{u \in \mathrm{An}(v)} b_{vu} Z_j, i = 1, \ldots, d \Big).$$

Given the DAG $\mathscr{D}$ and the edge weights c_{ik} with $c_{ii} = 1$ for all $i = 1, \ldots, d$, the max-linear coefficient matrix B can be found by iterating the weighted adjacency matrix $C = (c_{vu} \mathbf{1}_{\mathrm{Pa}(v)}(u))_{d \times d}$ of $\mathscr{D}$ using this matrix multiplication; here, $\mathbf{1}_{\mathrm{Pa}(v)}$ denotes the indicator function of $\mathrm{Pa}(v) = \mathrm{pa}(v) \cup \{v\}$):

$$B = \bigvee_{k=0}^{d-1} C^{\odot k} = (I \vee C)^{\odot(d-1)}, \tag{6.9}$$

cf. Butkovič (2010) [Lemma 1.4.1]. For more details, see Gissibl and Klüppelberg (2018) [Thm. 2.4].

By (6.6), the max-linear coefficient b_{vu} of X is different from zero if and only if $u \in \mathrm{An}(v)$. This information is contained in the *reachability matrix* $R = (r_{vu})_{d \times d}$ of $\mathscr{D}$, which has entries

$$r_{vu} := \begin{cases} 1, & \text{if there is a path from } u \text{ to } v, \text{ or if } u = v, \\ 0, & \text{otherwise.} \end{cases}$$

If the vu-th entry of R is equal to one, then v *is reachable from* u. In the context of a DAG $\mathscr{D}$ with its reachability matrix R and a recursive max-linear model X on $\mathscr{D}$ with max-linear coefficient matrix B, it will be useful to keep the following in mind.

Remark 6.1 Let $\mathscr{D}$ be a DAG with reachability matrix R.

(i) The max-linear coefficient matrix B is a weighted reachability matrix of $\mathscr{D}$; that is, $R = \mathrm{sgn}(B)$.
(ii) Since V is assumed well-ordered, B and R are lower triangular matrices.

From (6.6) and (6.7), we conclude that a path π from u to v, whose weight $d_{vu}(\pi)$ is strictly less than b_{vu}, does not have any influence on X_i. For $v \in V$ and $u \in \mathrm{an}(v)$ we call a path π from u to v *max-weighted*, if $b_{vu} = d_{vu}(\pi)$, and investigate its relevance for the recursive max-linear model in further detail.

Firstly, we note that we can remove an edge from $\mathscr{D}$ which is not part of a max-weighted path without changing the distribution of X. The DAG obtained in this way is termed the *minimum max-linear* DAG $\mathscr{D}^B$. In the special case where $\mathscr{D}$ is a polytree, all paths are necessarily max-weighted and we clearly have

Proposition 6.3 *If $\mathscr{D}$ is a polytree, it holds that $\mathscr{D}^B = \mathscr{D}$.*

The following result describes exactly all DAGs and edge weights possible for a given max-linear coefficient matrix. Recall that we have set $c_{vv} = 1$.

Theorem 6.3 *(Gissibl and Klüppelberg 2018, Thm. 5.4)*

Let X be given by a recursive max-linear structural equation system with coefficient matrix B. Let further $\mathscr{D}^B = (V, E^B)$ be the minimum max-linear DAG of X and $\mathrm{pa}^B(v)$ the parents of v in $\mathscr{D}^B$.

(a) *$\mathscr{D}^B$ is the DAG with the minimum number of edges such that X satisfies (6.4). The weights in (6.4) are uniquely given by $c_{vv} = b_{vv}$ and $c_{vs} = b_{vs}$ for $v \in V$ and $s \in \mathrm{pa}^B(v)$.*

(b) *Every DAG with vertex set V that has at least the edges of $\mathscr{D}^B$ and the same reachability matrix as $\mathscr{D}^B$ represents X in the sense of (6.4) with weights satisfying*

$$c_{vv} = b_{vv}, \ c_{vs} = b_{vs} \ \text{for } s \in \mathrm{pa}^B(v), \ \text{and } c_{vs} \in (0, b_{vs})$$

$$\text{for } s \in \mathrm{pa}(v) \setminus \mathrm{pa}^B(v).$$

There are no further DAGs and weights such that X has representation (6.4).

In general, recursive max-linear models are not faithful to their DAG, not even if $\mathscr{D} = \mathscr{D}^B$, see Remark 3.9 (ii) in Gissibl and Klüppelberg (2018). This is illustrated in Example 6.5 below.

Example 6.5 The example has been given in Example 6.8 of Gissibl and Klüppelberg (2018) and considers the same graph as Example 6.1.

We note that the paths $[1 \to 2]$, $[1 \to 3]$, $[2 \to 4]$, and $[3 \to 4]$ are max-weighted as they are the only directed paths between their endpoints. It therefore holds that $\mathscr{D}^B = \mathscr{D}$ since they are the unique max-weighted paths. Still, the distribution determined by this recursive system is never faithful to $\mathscr{D}$, as we shall see below.

Concerning the paths from node 1 to 4, we have three situations:

$$c_{42}c_{21} = c_{43}c_{31}, \quad c_{42}c_{21} > c_{43}c_{31}, \quad \text{or} \quad c_{42}c_{21} < c_{43}c_{31}.$$

In the first situation, both paths from 1 to 4, $[1 \to 2 \to 4]$ and $[1 \to 3 \to 4]$, are max-weighted whereas in the other situations only one of them is.

If the path $[1 \to 2 \to 4]$ is max-weighted, we can consider the subdag $\tilde{\mathscr{D}}$ obtained from $\mathscr{D}$ by removing the edge $1 \to 3$:

In other words, we are changing the edge weights by letting $\tilde{c}_{31} = 0$, keeping the other edge weights unchanged. The new max-linear coefficient matrix becomes

$$\tilde{B} = \begin{pmatrix} 1 & 0 & 0 & 0 \\ c_{21} & 1 & 0 & 0 \\ 0 & 0 & 1 & 0 \\ c_{42}c_{21} & c_{42} & c_{43} & 1 \end{pmatrix}$$

where we have exploited that $c_{ii} = 1$. The max-linear coefficient matrix for the marginal distribution of (X_1, X_2, X_4) is obtained by ignoring the third row and since only entries in the third row have changed, we see that (X_1, X_2, X_4) has the same joint distribution in the model determined by $\mathscr{D}$ as it has in the model determined by $\tilde{\mathscr{D}}$.

But as we clearly have $1 \perp_{\tilde{\mathscr{D}}} 4 \,|\, 2$, we conclude that $X_1 \perp\!\!\!\perp X_4 \,|\, X_2$ in the model determined by $\tilde{\mathscr{D}}$ and hence also by $\mathscr{D}$. But since $\neg(1 \perp_{\mathscr{D}} 4 \,|\, 2)$, the distribution is not faithful to $\mathscr{D}$.

If $[1 \to 3 \to 4]$ is also max-weighted, the similar argument yields $X_1 \perp\!\!\!\perp X_4 \,|\, X_3$, so the distribution is *not faithful to* $\mathscr{D}$ for any allocation of edge weights.

We note that Gissibl and Klüppelberg (2018) suggest in their Remark 3.9(i) that additional conditional independence relations that are valid for a given DAG can be revealed by considering a system of submodels determined by appropriate subgraphs, but here we refrain from giving a complete description of all valid conditional independence relations.

6.4   Statistical Properties

The statistical theory of recursive max-linear models is challenging because standard assumptions for smooth statistical models are not satisfied. For example, if we for a given DAG $\mathscr{D}$ consider the family $\mathscr{P}$ of distributions with coefficients adapted to $\mathscr{D}$, this family is not dominated by any measure on the space of observations, so standard likelihood theory does not apply. On the other hand, as we shall see, estimation of coefficients and identification of the network structure for recursive max-linear models can be made in a simple fashion and procedures are more efficient than usual in that estimates of coefficients and structures converge at exponential rates to the true values. Here, we shall give a summary of the most important findings in Gissibl et al. (2019).

Throughout the following we consider a sample $\mathbf{x} = (X^1 = x^1, \ldots, X^n = x^n)$ from a distribution P given by the recursive max-linear model (6.4).

6.4.1   Estimation of Coefficients

We first consider the situation where the DAG $\mathscr{D} = (V, E)$, and for the sake of simplicity, we assume the distribution of noise variables $Z_v, v \in V$, is completely known, the coefficients c_{vv} are all equal to one, whereas the edge weights $C = \{c_{vu}, u \in \mathrm{pa}(v), v \in V\}$ are all strictly positive, but otherwise unknown. We let $\mathscr{C}$ denote the set of all possible coefficients and P_C denote the distribution of X determined by the corresponding recursive model (6.4).

The family $\mathscr{P} = P_C, C \in \mathscr{C}$, is not dominated by any fixed σ-finite measure μ on $\mathscr{X}$, as the support of P_C varies strongly with the coefficients; more precisely, the distributions have disjoint atomic components. This is a disadvantage in the sense that we cannot define a standard likelihood function; but, as we shall see, an advantage since these atomic components help identifying P_C from a given sample. We illustrate this by a simple example.

Example 6.6 Consider the simple DAG $1 \to 2$ with just two nodes and a single directed edge, and let $c = c_{21}$ be the corresponding coefficient. We estimate c from an atom.

Then, P_c has support on the cone given as $x_2 \geq cx_1 \geq 0$ and the line $A_c = \{x_2 = cx_1\}$ is an atom for P_c because $P_c(A_c) = P(Z_2 \leq cX_1) = P(Z_2 \leq cZ_1) > 0$.

Still, since then $\{c\}$ is the only atom in P_c for $Y = X_2/X_1$, the sample will for large n with high probability have repeated values of Y and c will be the only value that is repeated. In other words, $\hat{c} = \min\{y^v = x_2^v/x_1^v, v = 1, \ldots, n\}$ will be exactly equal to the true parameter with high probability. A similar estimator has been considered by Davis and Resnick (1989) in a time series framework.

Although most likelihood theory is concerned with dominated families, Kiefer and Wolfowitz (1956) considered the non-dominated case. Their formulation has been used rarely — an exception being Johansen (1978); see also Scholz (1980) and Gill et al. (1989), for example. This formulation turns out to be exactly what we need to discuss estimation of C in a formal way.

For two probability measures P and Q on a measurable space $(\mathscr{X}, \mathbb{E})$, we define the *generalized likelihood ratio* $\rho_x(P, Q)$ at the observation x as

$$\rho_x(P, Q) = \frac{\mathrm{d}P}{\mathrm{d}(P + Q)}(x) \tag{6.10}$$

where $\mathrm{d}P/\mathrm{d}(P + Q)$ is the density of P w.r.t. $P + Q$; the density always exists as, clearly, $P(A) + Q(A) = 0 \implies P(A) = 0$ so P is absolutely continuous w.r.t. $P + Q$.

The idea here is that if $\rho_x(P, Q) > \rho_x(Q, P)$, then P is a more likely explanation of x than Q. We note in particular that if P and Q have densities f and g w.r.t. a σ-finite measure μ, we have $\rho_x(P, Q) = f(x)/\{f(x) + g(x)\}$ so then $\rho_x(P, Q) > \rho_x(Q, P)$ if and only if $f(x) > g(x)$. Hence, ρ_x extends the standard likelihood ratio in a natural way.

Clearly, the generalized likelihood ratio suffers from the same problem as the usual likelihood ratio: the densities are only defined almost surely, so can be changed on $P + Q$-null sets; therefore, a version of $\mathrm{d}P/\mathrm{d}(P + Q)$ must be chosen independently of the observation x.

Next we say that if $\mathscr{P}$ is a family of probability distributions, $\hat{P}$ is a *generalized maximum likelihood estimate* (GMLE) of P based on $x \in \text{supp}(\hat{P})$ if

$$\rho_x(\hat{P}, Q) \geq \rho_x(Q, \hat{P}) \text{ for all } Q \in \mathscr{P},$$

i.e. if $\hat{P}$ explains x at least as well as any other member of $\mathscr{P}$.

Example 6.7 We illustrate use of the generalized maximum likelihood ratio for the model described in Example 6.6. To identify the density, we consider two values $c > c^*$ where we have

$$\rho_x(c, c^*) = \frac{\mathrm{d}P_c}{\mathrm{d}(P_c + P_{c^*})}(x_1, x_2) = \begin{cases} 1/2 & \text{for } x_2 > cx_1 \\ 1 & \text{for } x_2 = cx_1 \\ 0 & \text{for } x_2 < cx_1 \end{cases}$$

and

$$\rho_x(c^*, c) = \frac{\mathrm{d}P_{c^*}}{\mathrm{d}(P_c + P_{c^*})}(x_1, x_2) = \begin{cases} 1/2 & \text{for } x_2 > cx_1 \\ 0 & \text{for } x_2 = cx_1 \\ 1 & \text{for } cx_1 > x_2 \geq c^*x_1 \\ 0 & \text{for } x_2 < c^*x_1. \end{cases}$$

If $c = c^*$, we may let

$$\rho_x(c, c) = \rho_x(c, c^*) = \rho_x(c^*, c) = \frac{1}{2}\mathbf{1}_{\{x_2 \geq cx_1\}}.$$

Thus, if we consider a full sample, let $\hat{c} = \min\{y^\nu = x_2^\nu/x_1^\nu, \nu = 1, \ldots, n\}$ and $n_+(c, \mathbf{x}) = \#\{\nu : y^\nu > c\}$, we get:

$$\rho_{\mathbf{x}}(\hat{c}, c) = \prod_{\nu=1}^{n} \rho_{x^\nu}(\hat{c}, c) = \begin{cases} 0 & \text{if } c > \hat{c} \text{ and } c \in \{y^\nu, \nu = 1, \ldots, n\} \\ 2^{-n_+(c,\mathbf{x})} & \text{if } c > \hat{c} \text{ and } c \notin \{y^\nu, \nu = 1, \ldots, n\} \\ 2^{-n} & \text{if } c = \hat{c} \\ 2^{-n_+(\hat{c},\mathbf{x})} & \text{if } c < \hat{c}, \end{cases}$$

whereas

$$\rho_{\mathbf{x}}(c, \hat{c}) = \prod_{\nu=1}^{n} \rho_{x^\nu}(c, \hat{c}) = \begin{cases} 0 & \text{if } c > \hat{c} \\ 2^{-n} & \text{if } c = \hat{c} \\ 0 & \text{if } c < \hat{c}. \end{cases}$$

Clearly, $\rho_{\mathbf{x}}(\hat{c}, c) \geq \rho_{\mathbf{x}}(c, \hat{c})$ showing that $\hat{c}$ is the unique GMLE of c.

Indeed, *it holds in general for a recursive max-linear model that*

$$\hat{c}_{ij} = \bigwedge_{v=1}^{n} \frac{x_i^v}{x_j^v}, \quad i \in V, j \in \mathrm{pa}(i)$$

is a GMLE of the edge weights. We refer to Gissibl et al. (2019) for further details but should point out that in the general case, the GMLE is not unique. Since the distribution of X only depends on the edge weights through the max-linear coefficient matrix B, only B is uniquely estimable from a sample. We clearly have by (6.9) for the GMLE that

$$\hat{B} = B(\hat{C}) = \bigvee_{k=0}^{d-1} \hat{C}^{\odot k} = (I \vee \hat{C})^{\odot(d-1)}.$$

An alternative estimate of the max-linear coefficient matrix is given as

$$\tilde{b}_{ij} = \bigwedge_{v=1}^{n} \frac{x_i^v}{x_j^v}, \quad i \in V, j \in \mathrm{an}(i).$$

Although this estimate is also sensible and asymptotically consistent, it is less efficient than the GMLE as X_i^v/X_j^v only attends its minimum value when all noise variables on the path from j to i are smaller than $b_{ij}X_j^v$ for the same v, whereas the minima for the X_v^v/X_u^v on the path from j to i can be attained for different vs.

6.4.2 Identification of Structure

General methods for identifying the structure of DAG $\mathscr{D}$ from a sample are often based on an assumption of faithfulness, so that observed conditional independence relations can be translated back to the structure of the DAG since then any observed conditional independence must correspond to a separation in $\mathscr{D}$, see, for example, Spirtes et al. (2000). Also, as noted in Thm. 6.2, two DAGs can be different but still Markov equivalent and thus any method based on observed direct conditional independence relations cannot distinguish between DAGs that are Markov equivalent.

As shown in Example 6.5, faithfulness is violated for max-linear Bayesian networks whenever $\mathscr{D}$ is not a polytree. However, as we shall see below, the minimal DAG $\mathscr{D}^B$ of a max-linear Bayesian network can still be completely recovered from observations.

This fact conforms with recent developments where the recursive linear structural equation systems have been shown to be completely identifiable if the errors follow a non-Gaussian distribution (Shimizu et al. 2006) and it has been shown that the faithfulness assumption can be considerably weakened also in other situations (Peters and Bühlmann 2014; Spirtes and Zhang 2014).

To explain why the structure $\mathscr{D}^B$ is identifiable, we consider the statistics

$$Y_{ij} = X_i / X_j, \quad i, j \in V$$

and note that Y_{ij} has support $[b_{ij}, \infty)$ and an atom in b_{ij} if and only if $j \in \mathrm{an}(i)$. Using this property, one can show that the following estimate $\check{B}$ eventually identifies the max-linear coefficient matrix B.

$$\check{b}_{ij} = \begin{cases} \bigwedge\limits_{\nu=1}^{n} y_{ij}^{\nu} & \text{if minimum value is attained at least twice in the sample,} \\ 0 & \text{otherwise.} \end{cases}$$

Then, $\mathscr{D}^B$ is identifiable from B; we refer the reader to Gissibl et al. (2019) for further details.

6.5 Conclusion

We have reviewed basic elements of Bayesian networks based on recursive max-linear structural equations and some of their statistical properties. We conclude this article by pointing out some natural extensions of this work that we hope to address in the future.

Firstly, it would be of interest to have a simple and complete description of all independence properties which hold for a distribution determined by a recursive max-linear equation system, i.e. a global Markov property for max-linear Bayesian networks.

Secondly, it appears that a consequent use of algebraic theory (Butkovič 2010), based on the properties of the max-times semiring $\mathbb{S} = ([0, \infty], \vee, \cdot)$, would be able to simplify the theory of these models.

Finally, we should emphasize that the models heuristically can be seen as limiting cases of standard linear recursive models where error distributions have heavy tails, and therefore, the maximal element of any sum will almost completely dominate the sum; a rigorous study of this limiting process will enhance the understanding of this class of models.

Acknowledgements The authors have benefited from discussions with Nadine Gissibl and financial support from the Alexander von Humboldt Stiftung.

References

Asadi, P., Davison, A. C. & Engelke, S. (2015), 'Extremes on river networks', *Ann. Appl. Stat.* **9**(4), 2023–2050. https://doi.org/10.1214/15-AOAS863

Beirlant, J., Goegebeur, Y., Segers, J., Teugels, J., De Waal, D. & Ferro, C. (2006), *Statistics of Extremes: Theory and Applications*, Wiley, Chichester.

Bollen, K. A. (1989), *Structural Equations with Latent Variables*, Wiley, New York.

Buhl, S., Davis, R. A., Klüppelberg, C. & Steinkohl, C. (2016), 'Semiparametric estimation for isotropic max-stable space-time processes', *arXiv:1609.04967*.

Butkovič, P. (2010) , *Max-linear Systems: Theory and Algorithms*, Springer, London.

Davis, R. A., Klüppelberg, C. & Steinkohl, C. (2013), 'Statistical inference for max-stable processes in space and time', *Journal of the Royal Statistical Society: Series B (Statistical Methodology)* **75**(5), 791–819.

Davis, R. A. & Resnick, S. I. (1989), 'Basic properties and prediction of max-ARMA processes', *Advances in Applied Probability* **21**(4), 781–803.

Davison, A. C., Padoan, S. A. & Ribatet, M. (2012), 'Statistical modeling of spatial extremes', *Statist. Sci.* **27**(2), 161–186. https://doi.org/10.1214/11-STS376

Dawid, A. P. (1980), 'Conditional independence for statistical operations', *The Annals of Statistics* **8**(3), 598–617.

de Haan, L. & Ferreira, A. (2006), *Extreme Value Theory: An Introduction*, Springer, New York.

Einmahl, J. H. J., Kirillouk, A. & Segers, J. (2018), 'A continuous updating weighted least squares estimator of tail dependence in high dimensions', *Extremes* **21**(2), 205–233. https://doi.org/10.1007/s10687-017-0303-7

Embrechts, P., Klüppelberg, C. & Mikosch, T. (1997), *Modelling Extremal Events: for Insurance and Finance*, Springer-Verlag Berlin Heidelberg.

Finkenstädt, B. & Rootzén, H. (2004), *Extreme Values in Finance, Telecommunications, and the Environment*, Chapman & Hall/CRC, Boca Raton.

Frydenberg, M. (1990), 'The chain graph Markov property', *Scandinavian Journal of Statistics* **17**(4), 333–353.

Gill, R. D., Wellner, J. A. & Præstgaard, J. (1989) , 'Non- and semi-parametric maximum likelihood estimators and the von Mises method (part 1)', *Scandinavian Journal of Statistics* **16**(2), 97–128.

Gissibl, N. & Klüppelberg, C. (2018), 'Max-linear models on directed acyclic graphs', *Bernoulli* **24**(4A), 2693–2720. https://doi.org/10.3150/17-BEJ941

Gissibl, N., Klüppelberg, C. & Lauritzen, S. (2019), 'Identifiability and estimation of recursive max-linear models', *arXiv:1901.03556*

Gissibl, N., Klüppelberg, C. & Mager, J. (2017), Big data: Progress in automating extreme risk analysis, *in* W. Pietsch, J. Wernecke & M. Ott, eds, 'Berechenbarkeit der Welt? Philosophie und Wissenschaft im Zeitalter von Big Data', Springer Fachmedien Wiesbaden, Wiesbaden, pp. 171–189.

Hoef, J. M. V., Peterson, E. E. & Theobald, D. (2006), 'Spatial statistical models that use flow and stream distance', *Environmental and Ecological Statistics* **13**(4), 449–464.

Huser, R. & Davison, A. C. (2014), 'Space-time modelling of extreme events', *Journal of the Royal Statistical Society: Series B (Statistical Methodology)* **76**(2), 439–461.

Johansen, S. (1978) , 'The product limit estimator as maximum likelihood estimator', *Scandinavian Journal of Statistics* **5**(4), 195–199.

Kiefer, J. & Wolfowitz, J. (1956), 'Consistency of the maximum likelihood estimator in the presence of infinitely many incidental parameters', *The Annals of Mathematical Statistics* **27**(4), 887–906.

Lauritzen, S. & Sadeghi, K. (2018), 'Unifying Markov properties for graphical models', *Ann. Statist.* **46**(5), 2251–2278. https://doi.org/10.1214/17-AOS1618

Lauritzen, S. (1996), *Graphical Models*, Oxford University Press, Oxford.

Lauritzen, S. L., Dawid, A. P., Larsen, B. N. & Leimer, H.-G. (1990), 'Independence properties of directed Markov fields', *Networks* **20**(5), 491–505.

Meek, C. (1995), Strong completeness and faithfulness in Bayesian networks, *in* P. Besnard & S. Hanks, eds, 'Proceedings of the 11th Conference on Uncertainty in Artificial Intelligence', Morgan Kaufmann Publishers, San Francisco, pp. 411–418.

Peters, J. & Bühlmann, P. (2014), 'Identifiability of Gaussian structural equation models with equal error variances', *Biometrika* **101**(1), 219–228.

Resnick, S. I. (1987), *Extreme Values, Regular Variation, and Point Processes*, Springer, New York.

Resnick, S. I. (2007), *Heavy-Tail Phenomena: Probabilistic and Statistical Modeling*, Springer, New York.

Richardson, T. (2003), 'Markov properties for acyclic directed mixed graphs', *Scandinavian Journal of Statistics* **30**(1), 145–157.

Scholz, F. W. (1980) , 'Towards a unified definition of maximum likelihood', *The Canadian Journal of Statistics* **8**(2), 193–203.

Shimizu, S., Hoyer, P. O., Hyvärinen, A. & Kerminen, A. (2006) , 'A linear non-Gaussian acyclic model for causal discovery', *J. Mach. Learn. Res.* **7**, 2003–2030.

Spirtes, P., Glymour, C. & Scheines, R. (2000) , *Causation, Prediction, and Search*, 2nd edn, MIT press.

Spirtes, P. & Zhang, J. (2014) , 'A uniformly consistent estimator of causal effects under the k-triangle-faithfulness assumption', *Statist. Sci.* **29**(4), 662–678. https://doi.org/10.1214/13-STS429

Verma, T. & Pearl, J. (1990), Equivalence and synthesis of causal models, *in* L. N. K. P. Bonissone, M. Henrion & J. F. Lemmer, eds, 'Proceedings of the 6th Conference on Uncertainty in Artificial Intelligence', Amsterdam, pp. 255–270.

Wang, Y. & Stoev, S. A. (2011), 'Conditional sampling for spectrally discrete max-stable random fields', *Adv. in Appl. Probab.* **43**(2), 461–483.

West, D. (2001), *Introduction to Graph Theory*, Prentice Hall.

Chapter 7
Introduction to Network Inference in Genomics

Ernst C. Wit

Abstract The genome is the archetypical complex system: it is a finely tuned whole whose many parts, such as DNA, RNA and proteins, interact at various levels to execute intricate functions, such as repair, replication and adapting to the external environment. One particularly effective way of conceptualizing this complex system is by means of a network, in which the vertices describe the genomic components and the edges describe their physical or functional interactions. With the advent of modern high-throughput genomic measuring devices, such as microarrays, RNA-seq and other next generation sequencing tools, it has become possible to measure the vertices of the genomic system in real time. One central question is whether from these measurements it is possible to reconstruct the edges of the genomic network. This essay describes three modelling and inference strategies to answer this central biological question.

7.1 Introduction

Networks have become an important paradigm to describe genomic systems: from describing the physical, molecular interactions between proteins to the abstract interactions between functional genetic units, the vocabulary of networks has been adopted eagerly by biologists tasked with studying complex biological systems. For example, Corominas et al. (2014) define the concept of spliceform networks for translating genetic knowledge into a better understanding of human diseases, whereas Costanzo et al. (2016) argues that a global genetic interaction network highlights the functional organization of a cell and provides a resource for predicting gene and pathway function.

Within biostatistics, mathematical biology and, more recently, bioinformatics, there have been a number of modelling and inference procedures proposed to capture genetic networks. Traditionally, metabolic pathway analysis has been using ordinary

E. C. Wit (✉)
Institute of Computational Science, Università Della Svizzera Italiana,
Lugano, Switzerland
e-mail: wite@usi.ch

© Springer Nature Switzerland AG 2019
F. Biagini et al. (eds.), *Network Science*, https://doi.org/10.1007/978-3-030-26814-5_7

differential equation models, or simplifications thereof, such as flux balance analyses (Papoutsakis 1984). This involved typically small network representations of a number of intertwined genetic pathways. With the advent of high-throughput genomic analysis, Boolean network representations of the transcription process became popular (Akutsu et al. 1999). More recently, stochastic differential equation models (Purutçuoğlu and Wit 2008; Wilkinson 2006), graphical models (Vinciotti et al. 2016), Bayesian networks (Grzegorczyk and Husmeier 2011) and vector autoregressive models (Abegaz and Wit 2013) have entered the scene.

In this chapter, we aim to introduce the reader to the way networks are being used in the analysis of biological systems. In Sect. 7.2, we describe a number of ways on how to think about various genomic systems *as* networks. In Sect. 7.3, we connect those systems with mathematical network models *and* high-throughput genomic data by showing what kinds of inference strategies are available for analysing those processes.

7.2   What Are Genomic Networks?

The language of genomic networks can be used in various ways, although roughly speaking biologists use "genes" as the nodes, connected by edges, which stands for some type of "genetic interactions". This may seem obvious, but the devil is in the details and there are various ways in which this can be made precise. Below we will consider a number of genomic networks, that each uses the concept of network in a somewhat different way.

Mechanic genomic networks

First, and perhaps, the most basal form of a genomic network is the molecular interactions between DNA, RNA and proteins. The interactions in this view are the physical binding of proteins to each other and to DNA, whereby the molecular functionality of the resulting molecule changes and leads to further downstream changes. This cascade of molecular interactions is typically initiated by outside forces, such as sunlight in the case of a circadian clock, the lack of water leading to a stress response in plants or the intake of food leading to production of energy by our mitochondria.

Figure 7.1 is an example of this first type of genomic network. It shows a simplified version of the MAPK-Erk pathway, which is a chain of proteins that via physical interactions carries the signal from a receptor on the surface of a cell to the DNA in the nucleus. It is a ubiquitous pathway and appears in the cell of many organisms. Malfunctioning of the MAPK-Erk pathway in humans has been linked to uncontrolled cell growth, and therefore cancer (Downward 2003). Understanding the activation, inhibition and feedback mechanisms in this network is, therefore, an important goal, which has already led to various drug targets (Hilger et al. 2002).

This mechanistic view of a genomic network is highly localized. The interactions described are individual binding events within a cell. Because of this, the event boundary of the network is typically the cell wall.

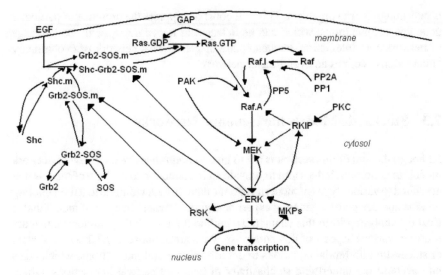

Fig. 7.1 Representation of the single-cell dynamics of the MAPK-Erk Network

Functional genomic networks

In contrast to the mechanistic description of a genomic network is a functional description. Although the nodes of this network can again be proteins or RNA, it is not uncommon that the nodes in this network are abstractly described as "genes". Typically, the focus is on larger systems than a single cell, such as organs or other biological subsystems. Interactions do typically not refer to specific mechanistic binding events, but rather to functional relationships. Often these networks are referred to as *gene regulatory networks.*

Just like the mechanistic genomic networks, the functional genomic network is most naturally interpreted as a dynamic process. However, whereas the changes in the mechanistic network are typically discrete, referring to a particular binding event, the functional network is more naturally seen as continuous, also referred to as a *flow network.*

Evolutionary networks

There are other genomic processes that can be described as a network, for example, how genes get passed on from generation to generation in the presence of genetic variability and selection. Most studied organisms are *diploid*, i.e. organisms that carry two copies of each gene. These copies can be the same, in which case we refer to them as *homozygous*, or different, in which case we refer to them as *heterozygous*. Mendel suggested that offspring receive a randomly selected version of each gene from either parent. Clearly, if the genetic make-up for a particular gene of the parents is the same and homozygous, then the offspring will be homozygous for that gene too. However, for many genetic loci within all species there is genetic variation, which means that offspring displays a "random" mosaic of the genetic make-up of their parents. Various constellations of this mosaic may lead to genetic advantage or disadvantage for the organism. This will boost or suppress the presence of particular genotype combinations, which can be detected as *dependence*, or in the language of networks: an interaction, between pairs of genes. This network of evolutionary "interactions" defines an evolutionary network.

7.3 Stochastic Models for Genomic Networks

Although the aim of this section is not to give a *comprehensive* overview of network models in genomics, it does aim to provide an introduction to the type of models that are suited to various types of modern genomic data. In fact, we argue that the sampling scheme and design of a genomic experiment should match the type of model that is used for analysing it. In this chapter, we outline three modelling strategies, that are useful in various aspects of this enterprise. We start in Subsect. 7.3.1 with a system of stochastic differential equations to describe single-cell interactions, which takes into account the underlying stochasticity of genomic particle interactions. Often, however, genomic data is collected at either a more agglomerated level or across a number of cells that are destructively sampled. In those cases, temporal models are more appropriately described by means of ordinary differential equations, described in Subsect. 7.3.2.1. In large genomic systems, both SDE and ODE descriptions can be unstable or computationally prohibitive. In such cases, vector autoregressive models, described in Subsect. 7.3.2.2, are useful. All these models are inherently dynamic. Nevertheless, the genotype is, at ordinary time-scales, a non-dynamic process, in which case it is more appropriate to describe these genomic interactions by means of a static network. This and other final considerations are described in Subsect. 7.3.3.

7.3.1 Modelling Mechanistic Genomic Networks

A cell is a natural unit of biology, whose state varies according to external influences and to internal regulation. The process of carrying over a signal, i.e. information, in the cell's environment is regulated by various signal transduction pathways. This signalling process is typically started by an external stimulus of the pathway leading to a binding of the signal to a receptor, i.e. hormones or growth factors, and ends by binding of a target protein. All cellular decisions such as cell proliferation, differentiation, or apoptosis are directed by different levels of transductions (Hornberg 2005). Deregulation of a single "renegade" cell can lead to diseases such as cancers, neurological disorders and developmental disorders (Macaulay et al. 2017).

Sequencing technologies now permit profiling the genome (Gawad et al. 2016), epigenome (Schwartzman and Tanay 2015), transcriptome (Stegle et al. 2015), or proteome (Wu and Singh 2012) of single cells sampled from heterogeneous cell types and cellular states. This allows us to study biological processes, such as disease development, at the cellular level. The technology is subject to measurement noise, but more importantly, the single-cellular process itself contains intrinsic stochasticity: the cellular system, characterized by its external environment and its internal protein levels, started at the same state may develop in different ways, merely by chance.

It is our aim to describe on the one hand the structured interactions between molecular particles and on the other hand the stochasticity involved in this process. We do this by means of a collection of random reaction equations. A general single-cellular, biochemical reaction can be defined as

$$k_1 Q_1 + k_2 Q_2 + \ldots + k_l Q_l \xrightarrow{\theta} s_1 P_1 + s_2 P_2 + \ldots + s_p P_p, \qquad (7.1)$$

where the terms on the left side, denoted as Q, are called the *reactants* and the ones on the right side, denoted as P, are named the *products*. The coefficients k_i ($i = 1, \ldots, l$) and s_j ($j = 1, \ldots, p$) represent the *stoichiometric coefficients* associated with the ith reactant Q_i and the jth product P_j, respectively. The quantity l refers the number of required reactants and p stands for the number of resulting products. So the chemical interpretation of this equation is that while molecules move around randomly in a cellular environment k_1 molecules of type Q_1, k_2 molecules of type Q_2, etc., "collide" with each other and produce s_1 molecules of type P_1, s_2 molecules of type P_2, etc. (Wilkinson 2006). Therefore under thermal equilibrium and fixed volume, a biochemical reaction shows which species and in what proportions react together and what they produce (Bower and Bolouri 2001).

For a set of r reactions and d species, accordingly, we can show the molecular transfer from reactant to product species as a net change of $V = S - K$ where V is called the $d \times r$ dimensional *net-effect* matrix when S denotes the $d \times r$ dimensional matrix of stoichiometry of products and K is the $d \times r$ dimensional matrix of stoichiometry of reactants. A reaction corresponds to a directed edge between the nodes $(Q_1, \ldots, Q_l)$ on the one hand and the nodes $(P_1, \ldots, P_p)$ on the other. The collection of r reactions, therefore, corresponds to a network with r directed edges between the d species or nodes of the network. This set of reactions can also contain uncertain, hypothesized reactions or even competing hypotheses, as shown in Fig. 7.2. This network is a representation of the potential stoichiometry between three proteins. The inference procedure with sufficient amount of data will eventually assign a zero reaction rate θ to reactions that are not part of the true underlying system. For example, if the reaction rate θ_1 associated with reaction 1 is inferred to be zero, then the resulting network would only involve the two reactions that are part of the second pathway. An over-parameterized system is, therefore, not a problem a priori and could be a modelling strategy to learn not only the kinetic parameters of the genomic system, but also the structure of the system.

We collect the amount of d reactants and products at time t in the vector U_t. They are put together in the same vector because products of one reaction are the reactants of another. There is therefore no fundamental difference between reactants and products. In the genomic context, they are typically proteins, protein complexes, enzymes, RNA and DNA. The aim is to define a probabilistic model for the evolution of the temporal process $\{U_t\}_t$. This is done by means of the master equation.

Fig. 7.2 The stochastic
differential equation models
could include competing
hypotheses. The data would
eventually weed out the links
for which there is no
evidence

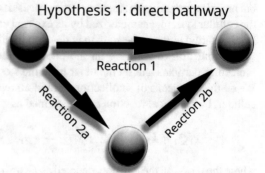

The master equation is defined as a differential equation for the process transition
probability and is written as:

$$\frac{dP(U;t)}{dt} = \sum_{k=1}^{r} \{h_k(U - V_{\cdot k}, \boldsymbol{\theta}) P(U - V_{\cdot k}, t) - h_k(U, \boldsymbol{\theta}) P(U, t)\}. \quad (7.2)$$

In other words, the probability of being in state U_t is positively related to the
tendency of the r available reactions to transit to state U_t and negatively related to
these same reactions to leave state U_t. The hazard h_k is a deterministic function of
the state and the reaction rate θ_k. For example, the reaction

$$2H + O \xrightarrow{\theta} H_2O,$$

in a volume with 5 hydrogen molecules H, 4 oxygen molecules and a rate of $\theta = 2$
reactions per time unit would lead to a hazard $h((5, 4), 2) = \binom{5}{2}\binom{4}{1}2 = 80$. By means
of a multivariate Taylor expansion, it is possible to derive an equivalent and alternative
formulation of any master equation, named the Kramers–Moyal expansion (Van
Kampen 1981):

$$\frac{dP(U;t)}{dt} = \sum_{m=1}^{\infty} \frac{(-1)^m}{m!} \sum_{j_1,\dots,j_m=1}^{N} \frac{d^m}{dU_{j_1},\dots,dU_{j_m}} [a_m(U, \boldsymbol{\theta}) P(U, t)], \quad (7.3)$$

where $a_m(U)$ are m-order symmetric tensors commonly called *jump moments* (Moyal
1949) or *propagator moment functions* (Gillespie 1992).

Various approximations to the process are possible. We can expand the distribution $P(U, t)$ by a second-order Taylor expansion and use a Fokker–Planck approach
for the change of each state (Bower and Bolouri 2001; Van Kampen 1981). This

stochastic expression is solved via Itô or Stratonovich integrals (Gillespie 1996; Golightly and Wilkinson 2005; Risken 1984; Van Kampen 1981) to obtain the following diffusion approximation

$$dU(t) = \mu(U_t, \theta)dt + \beta^{\frac{1}{2}}(U_t, \theta)dW(t), \tag{7.4}$$

where

$$\mu(U, \theta) = V'h(U, \theta),$$
$$\beta(U, \theta) = V'\text{diag}\{h(U, \theta)\}V$$

are the *drift* and *diffusion* matrices, respectively, both explicitly depending on state $U_t = (U_{t1}, \ldots, U_{td})$ at time t, the parameter vector $\theta = (\theta_1, \theta_2, \ldots, \theta_r)'$ and the net-effect matrix V. The expression $dW(t)$ represents the change of a Brownian motion during the time interval dt and $dU(t)$ shows the change in state U over time dt. This is effectively a large volume approximation that follows from the central limit theorem, whereby the reactions follow a Poisson process with rate $h(U, \theta)$ and the states changes therefore have mean $V'h(U, \theta)$ and variance $V'\text{diag}\{h(U, \theta)\}V$.

Due to the difficulties of inference of continuous-time multivariate diffusions, a further discrete Euler–Maruyama approximation is possible,

$$\Delta U_t = \mu(U_t, \theta)\Delta t + \beta^{\frac{1}{2}}(U_t, \theta)\Delta W_t \tag{7.5}$$

where ΔU_t is the change of state U over small time interval $[t, t + \Delta t]$ and ΔW_t is a d-dimensional independent identically distributed Gaussian random vector $\Delta W_t \sim N(0, I\Delta t)$ (Eraker 2001).

Data
The genomic interactions described above form a continuous-time process $\{U_t\}_t$ of gene activities on top of a genomic network. At best, we will be able to see snapshots X_t from this process. We will assume that we will have discrete observations $\{X_t\}_t$ from a single-cell genomic system $\{U_t\}_t$. For simplicity of presentation, we assume that the observations are equally spaced at regular time intervals of steps of size $\Delta t = 1$. This is merely for notational simplicity and not important for the inferential methods we use. There may be two types of missing values: first of all, several substrates may not be observed. It is quite common that due to technological limitations or experimental errors, it is not possible to measure the activity of all genomic species of interest. Various experimental techniques, such as microarrays, Chip-Seq analysis or mass-spectroscopy, have limitations to what they can measure. Furthermore, as most current technologies are capable of only discrete snapshots, the non-observed time points can also be considered missing.

Inference
There are various approaches possible for inference in such systems. The main issue the methods need to deal with is that the rate of change of the process is typically

faster than the observation rate, which leads to nonlinearities between the observation times. Frequentist approaches typically rely on the conditional nonlinear first and second moments of the process to propose a method of moments estimator for the reaction rates. To define a method of moment estimator via a generalized least squares objective function that can be minimized in order to estimate the unknown parameters vector θ:

$$\hat{\theta} = \arg \min_{\theta} \; (X_{1:T} - m(\theta))' \, W^{-1} \, (X_{1:T} - m(\theta)) \qquad s.t. \, \theta \geq 0_r$$

where

$$X_{1:T} = \begin{bmatrix} X_1 \\ X_2 \\ \vdots \\ X_T \end{bmatrix} \text{ and } m(\theta) = \begin{bmatrix} m(1; \theta) \\ m(2; \theta) \\ \vdots \\ m(T; \theta) \end{bmatrix}$$

are dT-dimensional column vectors with the observed cell-type count data and predicted mean evolutions, respectively. (Sotiropoulos and Kaznessis 2011) provide a general schema to derive analytical expressions for jump moments for any Markov process. Furthermore,

$$W = \begin{bmatrix} b(X_0; \theta) & 0 & \cdots & 0 \\ 0 & b(X_1; \theta) & \cdots & 0 \\ \vdots & \vdots & \ddots & \vdots \\ 0 & 0 & \cdots & b(X_{T-1}; \theta) \end{bmatrix}.$$

is a $dT \times dT$ block diagonal matrix, in which blocks correspond to expected variance-covariance matrices and zeros reflect the independence among measurements belonging to different time points.

An alternative way to deal with partially observed process is defining an augmented state space in combination with Bayesian inference. By inserting intermediate, unobserved states, the process can be linearized in the augmented, latent space. In a Bayesian approach to infer the kinetic parameters θ of the stochastic differential equation, one can use MCMC inference for calculating the posterior of the Euler–Maruyama system described in (7.5). Typically, Gibbs sampling can be difficult, because of the above-described data sparsity. In principle, it is possible to augment the data X with "missing" observations Z. A large number of augmented states in the Bayesian method increases the precision of the Euler–Maruyama approximation, but deteriorates the mixing of the Markov chain. Additional details about this problem and suggested solutions can be found in (Roberts and Stramer 2001) and (Golightly and Wilkinson 2008). In order to deal with these types of missingness, one can use a *Metropolis-within-Gibbs* step (Carlin and Louis 2000), whereby a Metropolis-Hastings step is implemented at each Gibbs step of the update. Therefore, the augmented process $U = \{U_t\}_{t=1}^{T}$ is a combination of X and Z, i.e. $U = (X, Z)$.

This method has been applied to estimating the MAPK-Erk pathway, consisting of 35 measured proteins and 16 unmeasured proteins across 77 time points that are involved in 66 reactions. Part of the inferred system is shown in Fig. 7.1. (Purutçuoğlu and Wit 2008) describe the biological interpretation of the results.

7.3.2 Modelling Functional Genomic Networks

Single-cell data, especially longitudinal single-cell data, are not very common. In fact, more often time-course genomic data are measured across a collection of cells. Moreover, not infrequently the measurements at different time points are on physically different samples. For example, various petri dishes with cells from some cell line are treated at a nominal time zero, and at various time points, the various dishes, one by one, are measured on the expression of their genomic constituents. As in many cases sampling tends to be destructive, each petri dish can be only measured once. This can be seen as cross-sectional sampling, where time is considered the factor of interest.

7.3.2.1 Ordinary Differential Equation Models

In such cases, it is not sensible to consider the stochastic relatedness between the various time points. However, it can still be interesting to consider the average dynamic behaviour of a genomic system. In fact, consider a simple reversible reaction,

$$A + B \xrightarrow{\theta_b, \theta_f} C,$$

where proteins A and B bind with forward rate θ_f into protein complex C, and, reversely, protein C breaks apart into constituents A and B with backward rate θ_b. According to the *Law of Mass Action* (Érdi and Tóth 1989), the average change in the amount of substrate A at time t_0 is negative proportional to the number of times forward reactions can happen, i.e. $a \times b$, and positively proportional to the number of times backward reactions can happen, i.e. c, where $A_t = a$, $B_t = b$ and $C_t = c$. This leads to the simple expression for the average change in A,

$$\frac{dm_A(t, \theta)}{dt} = c\theta_b - ab\theta_f.$$

Similarly, for B and C we have,

$$\frac{dm_B(t, \theta)}{dt} = c\theta_b - ab\theta_f,$$

$$\frac{dm_C(t, \theta)}{dt} = ab\theta_f + c\theta_b.$$

However, whereas the *Law of Mass Action* suggests a linear increase in the production rate of the product with an increase of the underlying substrate, in practice the increase will saturate. One reason is that there is only a finite amount of enzymes available, which are crucial auxiliary components in the genomic transcription system. (Michaelis and Menten 1913) introduced an intermediate substrate–enzyme complex, $C = SE$, in the transcriptional system,

$$S + E \overset{\theta_b, \theta_{f_1}}{\longleftrightarrow} C \overset{\theta_{f_2}}{\longleftrightarrow} P + E.$$

Combining the assumption of a finite amount of enzyme, $C + E =$ constant, with a *mass action* equilibrium $(\theta_b + \theta_{f_2})C_t = \theta_{f_1} S_t E_t$, they derived the so-called nonlinear Michaelis–Menten kinetics,

$$\frac{dm_P(t, \theta)}{dt} = \frac{\theta_{f_2} s}{\frac{\theta_b + \theta_{f_2}}{\theta_{f_1}} + s}.$$

Fig. 7.3 shows the typical saturation effect of the Michaelis–Menten production rate. This shows that for realistic descriptions of genomic interactions, we may have to consider a wider class of functions beyond *mass action* kinetics.

For the purposes of this overview, we will focus on a class of nonlinear ODEs that are linear in the rate parameters. Any of the models satisfying the *Law of Mass Action* satisfy also this requirement, but the class is larger than that and can accommodate saturation effects. Consider the gene regulatory or signalling network, described by a system of ordinary differential equations of the form

$$\begin{cases} z'(t) = g(z(t))\theta \text{ for } t \in [0, T], \\ z(0) = \xi, \end{cases} \tag{7.6}$$

Fig. 7.3 Typical saturation of the Michaelis–Menten production rate for various choices of the kinetic parameters

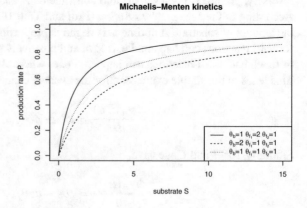

where $x(t)$ takes values in $\mathbb{R}^d$, with, possibly unknown, initial values $\xi \in \mathbb{R}^d$, and with the parameters of interest unknown $\theta \in \mathbb{R}^p$. We assume that $g = (g_1, \ldots, g_d)'$ is a known function, whose components $g_j : \mathbb{R}^d \to \mathbb{R}^p$. In particular, we consider a special case in which we want to model the change of each substrate by a saturating function of all the other substrates, i.e.

$$\begin{cases} z'_j(t) = \sum_{k=1}^{d} \theta_{kj} \log(z_k(t) + 1), \\ z_j(0) = \xi_j \end{cases} \quad j = 1, \ldots, d. \quad (7.7)$$

This model defines a network between the d substrates in that if $\theta_{jk} \neq 0$, then substrate k affects the change in substrate j. The logarithmic function is chosen to deal with natural saturation effects. Moreover, by its very definition $z_k(t) \geq 0$ and the leading Taylor term of $\log(z + 1)$ near zero is z, similar to the Michaelis–Menten production term. The solution $z(\cdot, \theta, \xi)$ implied by the ODE (7.7) — or more generally (7.6) — is assumed to be the mean of the observations taken from the system. In particular, we assume that at time points $t_i \in [0, T]$, $i = 1, \ldots, n$, we observe

$$X_j(t_i) = z_j(t_i, \theta, \xi) + \varepsilon_j(t_i), \quad j = 1, \ldots, d_1; i = 1, \ldots, n, \quad (7.8)$$

where $0 \leq t_1 < \cdots < t_n = T < \infty$ and $\varepsilon_i(t_j)$ is the measurement error for x_i at time t_j. The problem is to estimate θ, and thereby the underlying gene regulatory network, from the data $\{X_j(t_i)\}_{ij}$.

Inference of ODE networks
Inference of parameters in ODEs is not straightforward due to the possibly computationally prohibitive calculation of ODE solution $z(\cdot, \theta, \xi)$ for lots of values of θ and ξ. Regularization-based approaches, which make use of properties of differential operators, have been proposed to avoid numerical integration of the system of differential equations (González et al. 2013, 2014; Steinke and Schölkopf 2008). In most cases, the main computational bottleneck lies in the optimization of a nonlinear objective function. Alternatively, the idea of smoothing can be used as a way to avoid numerical integration of the system of differential equations and is referred to as the *collocation* estimation method; for example, there are *two-step* methods (Bellman and Roth 1971; Brunel 2008; Dattner and Klaassen 2013; Fang et al. 2011; Gugushvili and Klaassen 2012; Gugushvili and Spreij 2012; Liang and Wu 2008; Varah 1982) and *generalized profiling* methods (Ramsay et al.2007; Qi and Zhao 2010; Xun et al. 2011; Hooker et al. 2013).

The method we present here is a special case of generalized Tikhonov regularization (Vujačić et al. 2016) and without penalization has been shown to be $\sqrt{n}$-consistent (Vujačić et al. 2015). We consider estimators of the parameters θ and ξ that are obtained by minimizing the integral equation derived from (7.6),

$$L(\xi, \theta) = \int_0^T \left\| z(t) - \xi - \int_0^t g(z(s)) \, ds \, \theta \right\|^2 dt, \quad (7.9)$$

with respect to ξ and θ, where $z(t) = (t; \theta, \xi)$ will be replaced by a suitable estimator. We divide the interval $[0, T]$ in $\lfloor\sqrt{n}\rfloor$ subintervals, so that in every interval, we have at least $\lfloor\sqrt{n}\rfloor$ observations in it. Let $S_i = [a_{i-1}, a_i)$ be the ith subinterval $i = 1, \ldots, \lfloor\sqrt{n}\rfloor - 1$ and $S_{\lfloor\sqrt{n}\rfloor} = [a_{\lfloor\sqrt{n}\rfloor-1}, a_{\lfloor\sqrt{n}\rfloor}]$ and let $S(t)$ denote the subinterval to which t belongs. The piecewise constant *window estimator* of z is defined as

$$\hat{z}(t) = \frac{1}{|S(t)|} \sum_{t_j \in S(t)} X(t_j), \qquad t \in S(t). \tag{7.10}$$

This estimator $\hat{z}(t)$ estimates $z(t)$ as the mean of the observations that belong to interval $S(t)$. This allows us to estimate the inner integral in (7.9),

$$G(t) = \int_0^t g(\hat{z}(s))ds$$
$$= \sum_{m=1}^{i-1} g(\hat{z}(S_m))(a_m - a_{m-1}) + g(\hat{z}(S_i))(t - a_{i-1}), \qquad \text{where } t \in S_i.$$

Throughout the paper, we adhere to the convention that the sums of the form $\sum_{m=1}^{i-1} f_m$ are equal to zero for $i = 1$. Minimizing the criterion function (7.9) with respect to $\omega = (\xi; \theta)'$ yields explicit formulas for the estimators of the parameters. Indeed, the objective function L can be written as a quadratic function of the parameters,

$$L(\omega) = \omega' \int_0^T F(t)'F(t)dt\omega - 2\omega' \int_0^T F(t)'\hat{z}(t)dt + \int_0^T \|\hat{z}(t)\|^2 dt,$$

where $F(t) = (T I_d; G(t))$. The minimizer of this quadratic expression is given by

$$\hat{\omega} = \left(\int_0^T F(t)'F(t)dt\right)^{-1} \int_0^T F(t)'\hat{z}(t)dt$$

which has an explicit form by means of finite sums as shown in (Vujačić et al. 2015). It can be shown that this estimator is $\sqrt{n}$-consistent.

Example 7.1 *Circadian clock in Arabidopsis*

Consider the previously introduced, heavily parameterized ODE describing the change of each substrate in the gene regulatory network by a slowly saturating function of all the other substrates, i.e.

$$\begin{cases} z'_j(t) = \sum_{k=1}^d \theta_{kj} \log(z_k(t) + 1), \\ z_j(0) = \xi_j \end{cases} \qquad j = 1, \ldots, d, \tag{7.11}$$

or using some other production terms, such as $g(z) = \sqrt{z}$ or simply $g(z) = z$. This relatively simple gene regulatory network contains d^2 interaction parameters $\theta =$

$\{\theta_{kj}\}$. Many of these parameters can be expected to be zero as only a few genes will be responsible for activating other genes.

To enforce sparsity, we will add a L_1 regularization term on the objective function (7.9),

$$L_\lambda(\theta, \xi) = L(\theta, \xi) + \lambda||\theta||_1.$$

The estimator of θ and ξ will depend on the tuning parameter λ. In fact, the path estimator $(\hat{\xi}_\lambda, \hat{\theta}_\lambda)$ will correspond to the original lasso estimator $\hat{\beta}_\lambda$ for a quadratic problem (Tibshirani 1996),

$$\hat{\beta}_\lambda = \arg\min_\beta (y - X\beta)'(y - X\beta) + \lambda||\beta||_1,$$

whereby $X'X = \int_0^T F(t)'F(t)\mathrm{d}t$ and $X'y = \int_0^T F(t)'\hat{z}(t)\mathrm{d}t$, whereby the first d parameters, corresponding to ξ, will not be penalized and always included in the solution path.

We illustrate our proposed approach by applying it to a time-course gene expression dataset related to the study of circadian regulation in plants. The data used in our study come from the EU project TiMet (FP7-245143, 2014), whose objective is the elucidation of the interaction between circadian regulation and metabolism in plants.

The data consist of transcription profiles for 9 core clock genes from the leafs of various genetic variants of *Arabidopsis thaliana*. The plants were grown in 3 light conditions: a diurnal cycle with 12-hour light and 12-hour darkness (12L/12D), an extended night with full darkness for 24 hours, and an extended light with constant light for 24 hours. Samples were taken every 2 hours to measure mRNA concentrations. In total, there are 51 measurements across time. The nine genes are known to be involved in circadian regulation (Grzegorczyk et al. 2008; Aderhold et al. 2014). They consist of two groups of genes: "Morning genes", which are LHY, CCA1, PRR9 and PRR5, whose expression peaks in the morning, and "Evening genes", including TOC1, ELF4, ELF3, GI and PRR3, whose expression peaks in the evening. The expressions for all the genes are strictly positive.

Figure 7.4 shows the resulting sparse ODE network inferred with three different functions g, two of which deal explicitly with possible saturation effects, such as $g(x) = \log(x + 1)$ and $g(x) = \sqrt{x}$ and the naive linear production function $g(x) = x$. The results are quite robust, but suggest that it is worth considering possible saturation effects.

7.3.2.2 Vector Autoregressive Models

Both SDE and ODE models are in principle generative models for the underlying process of interest. Their aim is to describe the intrinsic relationship between the genomic substrates, typically on the basis of the *Law of Mass Action* or extensions thereof. Often, part of the model is inspired by biological knowledge. In this section,

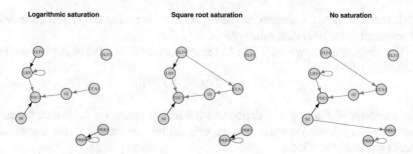

Fig. 7.4 Circadian clock network in *Arabidopsis thaliana*: red arrows represent suppression, whereas black arrow suggests activation. The ODE network inference results are quite robust, whether one considers a saturation model, whereby the effect on the production term depends on $g(z) = \log(z + 1)$ or $g(z) = \sqrt{z}$, or one that does not saturate, in which case $g(z) = z$. The yellow genes are the morning genes, whereas the blue genes are the evening genes

we describe a method fundamentally aimed at a more exploratory approach of high-dimensional genomic time series data. The idea is to explore potential temporal interactions between substrates, without focusing on the details of the kinetics. For this, we will use vector autoregressive models (VARs), which have been studied more in detail in the econometric literature (Dahlhaus and Eichler 2003). The details of the method described in this section can be found in (Abegaz and Wit 2013).

Within a vector autoregressive model, the time-course gene–gene interactions evolve according to Markovian dynamics, rather than an explicit functional form as in the ODE approach. Specifically, within a VAR(1) model the vector of gene expressions at time t relates only to those at time $t - 1$; extensions to a Markovian lag dependence greater than 1 are straightforward. Let X_t be a d-dimensional random vector associated with the expression of the d genes at time t. According to the first-order Markov property, the joint probability density of $X_0, \ldots, X_T$ can be decomposed as:

$$f(X_0, \ldots, X_T) = f(X_0)f(X_1 \mid X_0) \times \cdots \times f(X_T \mid X_{T-1}). \qquad (7.12)$$

We focus only on the conditional distributions in (7.12) and ignore the initial term $f(X_0)$. Furthermore, we assume a time-homogeneous dynamic network structure for the conditional distribution $f(X_t \mid X_{t-1})$ that can be approximated via a multivariate Gaussian,

$$X_t \mid X_{t-1} \sim N(\Gamma X_{t-1}, \Sigma). \qquad (7.13)$$

This vector autoregressive process of order one can also be expressed as

$$X_t = \Gamma X_{t-1} + \epsilon_t, \qquad (7.14)$$

where $\epsilon_t \sim N(0, \Sigma)$. The parameter elements in the matrices Γ and in the inverse of Σ represent directed and undirected links in the Markovian conditional independence graph, respectively. In particular, a nonzero element in Γ, say $\Gamma_{ij} \neq 0$, corresponds to a directed edge in the conditional independence graph between gene j at the previous time point and gene i at the current one. This edge is given the name *Granger causality* and reflects a delayed interaction between two genes, which can be cautiously given a semi-causal interpretation (Granger 1988). Given Σ and the corresponding precision matrix $\Theta = \Sigma^{-1}$ undirected edges relate to nonzero elements in the precision matrix Θ. If $\Theta_{ij} \neq 0$, then after adjusting for the past and present effects of other genes, there is an instantaneous interaction, or dependence, between genes i and j. A cartoon representation of the model formulation is given in Fig. 7.5.

Data

Suppose that we have n replications of a T time point longitudinal microarray study across p genes. The data, then, can be summarized as an $n \times p \times T$ array $X = (X_1, \ldots, X_n)'$ whose ith submatrix X_i has columns such that $X_{i.t} = (X_{i1t}, \ldots, X_{ipt})'$ which correspond to the expression levels of p genes measured at time t. That is, X_{ijt} is the jth gene expression level at time t for the ith replicate.

Sparse VAR network inference

The inference aim is to reconstruct the dynamic and contemporaneous genomic networks. Time-course genomic data typically consist of hundreds or thousands of genes measured on a comparatively small number of replications (typically 3) of microarray experiments across a few time steps (often not more than 10). The model formulation in (7.14) is in a standard vector autoregressive form with correlated errors and estimation approach for high-dimensional time-course genomic data is challenging. (Abegaz and Wit 2013) proposes a penalized maximum likelihood estimation methods for the analysis of the high-dimensional time-course gene expression data. The

Fig. 7.5 The dynamic network encoded in Γ shows that gene 1 is an important regulator. The instantaneous network Θ shows a central role of gene 2, but because the genomic interaction times are faster than the sampling times δt, it is not possible to say whether gene 2 regulates the other or the other way around

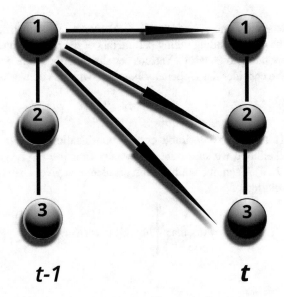

proposed approach provides sparse estimates of the autoregressive coefficient matrix Γ and the precision matrix Θ in (7.14), which are used to reconstruct the genomic network.

Under the Gaussian assumption described in (7.13), the conditional density of the tth observation is given by

$$f_c\left(X_t \mid X_{t-1}; \Gamma, \Theta\right) = (2\pi)^{p/2}|\Theta|^{1/2} \exp\left[-\frac{1}{2}\left(X_t - \Gamma X_{t-1}\right)' \Theta \left(X_t - \Gamma X_{t-1}\right)\right].$$

Then the conditional log-likelihood for n replicates each at T time steps becomes

$$\ell(\Gamma, \Theta) = \sum_{i=1}^{n} \sum_{t=1}^{T} \log f_c\left(X_{it} \mid X_{i,t-1}; \Gamma, \Theta\right)$$

$$= -\frac{npT}{2}\log(2\pi) + \frac{nT}{2}\log|\Theta| - \frac{nT}{2}tr(S_\Gamma\Theta), \qquad (7.15)$$

where

$$S_\Gamma = (1/nT) \sum_{i=1}^{n} \sum_{t=1}^{T} \left(X_{it} - \Gamma X_{i,t-1}\right)\left(X_{it} - \Gamma X_{i,t-1}\right)'.$$

We consider a penalized likelihood framework, where the objective function based on (7.15) is defined as

$$\ell_{pen}(\Gamma, \Theta) = \log|\Theta| - tr(S_\Gamma\Theta) - \sum_{i \neq j}^{p} P_\lambda(|\theta_{ij}|) - \sum_{i \neq j}^{p} P_\rho(|\gamma_{ij}|), \qquad (7.16)$$

where θ_{ij} and γ_{ij} are the (i, j)-elements of the matrix Θ and Γ and λ and ρ are the corresponding tuning parameters of the penalty functions $P_\lambda(\cdot)$ and $P_\rho(\cdot)$ corresponding to Θ and Γ. Various penalty functions have been proposed in the literature. We consider the L_1 penalty function, which is convex and given by

$$P_\lambda(\theta) = \lambda|\theta|, \qquad P_\rho(\gamma) = \rho|\gamma|. \qquad (7.17)$$

This leads to a desirable convex optimization problem. To obtain the L_1 penalized likelihood we substitute the penalty function in (7.17) into the objective function (7.16). Then, the optimization problem that gives sparse estimates of Γ and Θ is the solution of

$$(\widehat{\Theta}, \widehat{\Gamma})_{\lambda,\rho} = \arg\max_{\Theta, \Gamma}\left\{\log|\Theta| - tr(S_\Gamma\Theta) - \lambda\sum_{i \neq j}^{p}|\theta_{ij}| - \rho\sum_{i,j}^{p}|\gamma_{ij}|\right\}. \qquad (7.18)$$

Model selection

Under the penalized maximum likelihood framework for time series chain graphical models, the sparsity of the estimated precision matrix Θ and the autoregressive coefficient matrix Γ are controlled by the tuning parameters λ and ρ. The Bayesian information criterion can be used for selecting parsimonious parameter representations (Yin and Li 2011). The BIC is defined as

$$BIC(\lambda, \rho) = -nT \left\{ \log |\widehat{\Theta}_\lambda| - tr(S_{\widehat{\Gamma}_\rho} \widehat{\Theta}_\lambda) \right\} + \log(nT)(a_n/2 + b_n + p), \quad (7.19)$$

where p is the number of variables, a_n is the number of nonzero off-diagonal elements of $\widehat{\Theta}_\lambda$ and b_n is the number of nonzero elements of $\widehat{\Gamma}_\rho$. Thus, we select the values of λ and ρ that minimizes the criterion in (7.19). Here the minimization of $BIC(\lambda, \rho)$ with respect to λ and ρ is achieved by a grid search.

Example 7.2 *Mammary gland gene expression network*

We illustrate the proposed approach on the analysis of mammary gland gene expression time-course data from (Stein et al. 2004). In the mammary gland expression experiment, there are 12,488 probe sets representing approximately 8,600 genes. These probe sets are measured over 54 arrays of 3 replicates on each of 18 time points. We identified 30 genes that yield the best separation between the four developmental stages (virgin, pregnant, lactating, involution) using cluster analysis. We implemented the sparse VAR procedure in the R package `SparseTSCGM`. We apply the proposed VAR model to study the interaction between these crucial genes that trigger the transitions to the main developmental events in the mammary gland of mice. Fig. 7.6(a) shows the undirected links associated with Θ, related to instantaneous interactions among the genes and Fig. 7.6(b) displays the directed links that indicate Granger causality relations among the genes.

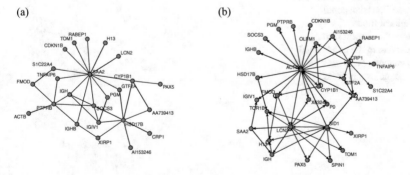

Fig. 7.6 Undirected (left) and directed (right) time series chain graphical model network inferred from the mammary gland time-course expression data with a VAR(1) model

7.3.3 Other Genomic Network Models

The models we have considered so far have all been dynamic network models. The main reason is that these models capture the dynamic nature of the genetic process. Depending on external stimuli and the internal state of a cell, The main reason is that these models capture the dynamic nature of the genetic process: at each moment the then relevant genes are transcribed, translated and broken down again in an intricate, interdependent process. Nevertheless, the three models that we have discussed are not the only ones that can be used. Some people might have noticed that we did not explicitly deal with Bayesian network models. Although they are closely related to vector autoregressive processes, the biostatistics and bioinformatics literature has seen many fine examples of such models applied to gene regulatory systems (Grzegorczyk and Husmeier 2011).

At the same time, there are also certain biological processes that can be modelled very elegantly by means of static network models. Genome-wide association studies (GWAS) are aimed at uncovering associations between genotype and phenotype. At the same time, certain genotype combinations might be evolutionary very advantageous or, more likely, detrimental. That is why such GWAS data can also be used to study epistasis by inferring the conditional independence graph: if there is no epistasis, the conditional independence graph will show the chromosomal backbone, whereas, if there is some epistasis, then we will find additional links between regions of the genome that are possibly on different chromosomes. Figure 7.7 shows such an example in *Arabidopsis thaliana*, which has been found by means of L_1 penalized Gaussian copula graphical modelling (Behrouzi and Wit 2017).

Fig. 7.7 Epistatic effects in a genotype study involving an *Arabidopsis thaliana* recombinant inbred line. The sparse Gaussian copula graphical model clearly shows the chromosomal backbone in the conditional dependency graph as a result of the meiosis process

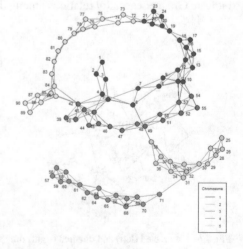

7.4 Discussion

In this chapter, we have looked at modelling dynamic biological networks. Unlike in social networks, this typically does not involve random graph models. The reason is that the biological phenomena of interest, such as gene transcription, pertain to the nodes of the network, rather than the edges. In other words, the random process of interest lives on the vertices of the graph. For this reason, the network models we have considered in this chapter are more closely connected to engineering networks used to describe flows.

Although networks have become an important modelling paradigm in genomics, there is currently no single network model to describe all the genomic interaction structures. In fact, it will be unlikely that there will ever be one. As the underlying generative model in biology is extremely complicated, we will always rely on convenient parameterizations to answer specific questions that arise in system biology. We have considered three types of models, namely stochastic differential equation models, ordinary differential equation models and vector autoregressive models and each of these modelling frameworks was selected depending on the underlying sampling design ("Are the measurements from a single cell or average over many cells?") and on the question of interest ("Do we want to describe the kinetics of the interactions or get an idea of the overall interaction structure of the genome?"). As George Box is said to have once said "all models are wrong, but some are useful" (Wit et al. 2012), and very useful indeed.

References

Abegaz, F. & Wit, E. (2013), 'Sparse time series chain graphical models for reconstructing genetic networks', *Biostatistics* **14**(3), 586–599.

Aderhold, A., Husmeier, D. & Grzegorczyk, M. (2014), 'Statistical inference of regulatory networks for circadian regulation', *Statistical applications in genetics and molecular biology* **13**(3), 227–273.

Akutsu, T., Miyano, S. & Kuhara, S. (1999), 'Identification of genetic networks from a small number of gene expression patterns under the Boolean network model', *Pacific Symposium on Biocomputing* pp. 17–28.

Behrouzi, P. & Wit, E. C. (2017), 'Detecting epistatic selection with partially observed genotype data by using copula graphical models', *Journal of the Royal Statistical Society: Series C (Applied Statistics)* .

Bellman, R. & Roth, R. S. (1971), 'The use of splines with unknown end points in the identification of systems', *Journal of Mathematical Analysis and Applications* **34**(1), 26–33.

Bower, J. M. & Bolouri, H. (2001), *Computational Modelling of Genetic and Biochemical Networks*, 2nd edn, Massachusetts Institute of Technology.

Brunel, N. J-B (2008), 'Parameter estimation of ODE's via nonparametric estimators', *Electronic Journal of Statistics* **2**, 1242–1267.

Carlin, B. P. & Louis, T. A. (2000), *Bayes and Empirical Bayes Methods for Data Analysis*, 2nd edn, Chapman and Hall/CRC.

Corominas, R., Yang, X., Lin, G. N., Kang, S., Shen, Y., Ghamsari, L., Broly, M., Rodriguez, M., Tam, S., Trigg, S. A. et al. (2014) , 'Protein interaction network of alternatively spliced isoforms from brain links genetic risk factors for autism', *Nature communications* **5**.

Costanzo, M., VanderSluis, B., Koch, E. N., Baryshnikova, A., Pons, C., Tan, G., Wang, W., Usaj, M., Hanchard, J., Lee, S. D. et al. (2016), 'A global genetic interaction network maps a wiring diagram of cellular function', *Science* **353**(6306), aaf1420.

Dahlhaus, R. & Eichler, M. (2003), Causality and graphical models in time series analysis, *in* R. S., ed., 'Highly Structured Stochastic Systems', Oxford University Press, pp. 115–137.

Dattner, I. & Klaassen, C. A. (2013), 'Estimation in systems of ordinary differential equations linear in the parameters', arXiv:1305.4126 .

Downward, J. (2003), 'Targeting RAS signalling pathways in cancer therapy', *Nature Reviews Cancer* **3**(1), 11.

Eraker, B. (2001), '$MCMC$ analysis of diffusion models with application to finance', *Journal of Business and Economic Statistics* **19**(2), 177–191.

Érdi, P. & Tóth, J. (1989), *Mathematical Models of Chemical Reactions: Theory and Applications of Deterministic and Stochastic Models*, Manchester University Press.

Fang, Y., Wu, H. & Zhu, L.-X. (2011), 'A two-stage estimation method for random coefficient differential equation models with application to longitudinal HIV dynamic data', *Statistica Sinica* **21**(3), 1145–1170.

Gawad, C., Koh, W. & Quake, S. R. (2016), 'Single-cell genome sequencing: current state of the science', *Nature reviews. Genetics* **17**(3), 175.

Gillespie, D. (1992), *Markov processes: An introduction for physical scientists.*, Academic Press.

Gillespie, D. T. (1996), 'The multivariate Langevin and Fokker-Planck equations', *American Journal of Physics* **64**(10), 1246–1257.

Golightly, A. & Wilkinson, D. J. (2005), 'Bayesian inference for stochastic kinetic models using a diffusion approximation', *Biometrics* **61**(3), 781–788.

Golightly, A. & Wilkinson, D. J. (2008), 'Bayesian inference for nonlinear multivariate diffusion models observed with error', *Computational Statistics and Data Analysis* **52**(3), 1674–1693.

González, J., Vujačić, I. & Wit, E. (2013), 'Inferring latent gene regulatory network kinetics', *Statistical applications in genetics and molecular biology* **12**(1), 109–127.

González, J., Vujačić, I. & Wit, E. (2014), 'Reproducing kernel Hilbert space based estimation of systems of ordinary differential equations', *Pattern Recognition Letters* **45**, 26–32.

Granger, C. W. (1988), 'Causality, cointegration, and control', *Journal of Economic Dynamics and Control* **12**(2-3), 551–559.

Grzegorczyk, M. & Husmeier, D. (2011), 'Improvements in the reconstruction of time-varying gene regulatory networks: dynamic programming and regularization by information sharing among genes', *Bioinformatics* **27**(5), 693–699.

Grzegorczyk, M., Husmeier, D., Edwards, K. D., Ghazal, P. & Millar, A. J. (2008), 'Modelling non-stationary gene regulatory processes with a non-homogeneous Bayesian network and the allocation sampler', *Bioinformatics* **24**(18), 2071–2078.

Gugushvili, S. & Klaassen, C. A. J. (2012), '$\sqrt{n}$-consistent parameter estimation for systems of ordinary differential equations: bypassing numerical integration via smoothing', *Bernoulli* **18**, 1061–1098.

Gugushvili, S. & Spreij, P. (2012), 'Parametric inference for stochastic differential equations: a smooth and match approach', *ALEA* **9**(2), 609–635.

Hilger, R., Scheulen, M. & Strumberg, D. (2002), 'The Ras-Raf-MEK-ERK pathway in the treatment of cancer', *Oncology Research and Treatment* **25**(6), 511–518.

Hooker, G., Ellner, S., Earn, D. et al. (2011), 'Parameterizing state-space models for infectious disease dynamics by generalized profiling: measles in ontario.', *Journal of the Royal Society, Interface* **8**(60), 961–974.

Hornberg, J. J. (2005), Towards integrative tumor cell biology control of MAP kinase signalling, PhD thesis, Vrije Universiteit, Amsterdam.

Liang, H. & Wu, H. (2008), 'Parameter estimation for differential equation models using a framework of measurement error in regression models', *Journal of the American Statistical Association* **103**(484), 1570–1583.

Macaulay, I. C., Ponting, C. P. & Voet, T. (2017), 'Single-cell multiomics: multiple measurements from single cells', *Trends in Genetics*.

Michaelis, L. & Menten, M. L. (1913), 'The kinetics of the inversion effect', *Biochem. Z* **49**, 333–369.

Moyal, J. (1949), 'Stochastic processes and statistical physics.', *Journal of the Royal Statistical Society. Series B* **11**, 150–210.

Papoutsakis, E. T. (1984), 'Equations and calculations for fermentations of butyric acid bacteria', *Biotechnology and bioengineering* **26**(2), 174–187.

Purutçuoğlu, V. & Wit, E. (2008), 'Bayesian inference for the MAPK/ERK pathway by considering the dependency of the kinetic parameters', *Bayesian Analysis* **3**(4), 851–886.

Qi, X. & Zhao, H. (2010), 'Asymptotic efficiency and finite-sample properties of the generalized profiling estimation of parameters in ordinary differential equations', *The Annals of Statistics* **38**(1), 435–481.

Ramsay, J. O., Hooker, G., Campbell, D. & Cao, J. (2007), 'Parameter estimation for differential equations: a generalized smoothing approach', *Journal of the Royal Statistical Society: Series B (Statistical Methodology)* **69**(5), 741–796.

Risken, H. (1984), *The Fokker-Planck Equation: Methods of Solution and Applications*, Springer-Verlag.

Roberts, G. O. & Stramer, O. (2001), 'On inference for partially observed nonlinear diffusion models using the Metropolis-Hastings algorithm', *Biometrika* **88**(3), 603–621.

Schwartzman, O. & Tanay, A. (2015), 'Single-cell epigenomics: techniques and emerging applications', *Nature reviews. Genetics* **16**(12), 716.

Sotiropoulos, V. & Kaznessis, Y. (2011), 'Analytical derivation of moment equations in stochastic chemical kinetics.', *Chemical engineering science* **66**(3), 268–277.

Stegle, O., Teichmann, S. A. & Marioni, J. C. (2015), 'Computational and analytical challenges in single-cell transcriptomics', *Nature reviews. Genetics* **16**(3), 133.

Stein, T., Morris, J. S., Davies, C. R., Weber-Hall, S. J., Duffy, M.-A., Heath, V. J., Bell, A. K., Ferrier, R. K., Sandilands, G. P. & Gusterson, B. A. (2004), 'Involution of the mouse mammary gland is associated with an immune cascade and an acute-phase response, involving lbp, cd14 and stat3', *Breast Cancer Res* **6**, R75–R91.

Steinke, F. & Schölkopf, B. (2008), 'Kernels, regularization and differential equations', *Pattern Recognition* **41**(11), 3271–3286.

Tibshirani, R. (1996), 'Regression shrinkage and selection via the lasso', *Journal of the Royal Statistical Society. Series B (Methodological)* pp. 267–288.

Van Kampen, N. G. (1981), *Stochastic Processes in Physics and Chemistry*, North-Holland, Amsterdam.

Varah, J. (1982), 'A spline least squares method for numerical parameter estimation in differential equations', *SIAM Journal on Scientific and Statistical Computing* **3**(1), 28–46.

Vinciotti, V., Augugliaro, L., Abbruzzo, A. & Wit, E. C. (2016), 'Model selection for factorial Gaussian graphical models with an application to dynamic regulatory networks', *Statistical applications in genetics and molecular biology* **15**(3), 193–212.

Vujačić, I., Dattner, I., González, J. & Wit, E. (2015), 'Time-course window estimator for ordinary differential equations linear in the parameters', *Statistics and Computing* **25**(6), 1057–1070.

Vujačić, I., Mahmoudi, S. M. & Wit, E. (2016), 'Generalized Tikhonov regularization in estimation of ordinary differential equations models', *Stat* **5**(1), 132–143.

Wilkinson, D. J. (2006), *Stochastic Modelling for Systems Biology*, Chapman and Hall/CRC.

Wit, E., Heuvel, E. v. d. & Romeijn, J.-W. (2012), "All models are wrong...': an introduction to model uncertainty', *Statistica Neerlandica* **66**(3), 217–236.

Wu, M. & Singh, A. K. (2012), 'Single-cell protein analysis', *Current Opinion in Biotechnology* **23**(1), 83–88.

Xun, X., Cao, J., Mallick, B., Maity, A. & Carroll, R. J. (2013), 'Parameter estimation of partial differential equation models', *Journal of the American Statistical Association* **108**(503), 1009–1020.

Yin, J. & Li, H. (2011), 'A sparse conditional Gaussian graphical model for analysis of genetical genomics data', *The Annals of Applied Statistics* **5**(4), 2630.

Printed in the United States
By Bookmasters